STUDENT SOLUTIONS MANUAL

to accompany

CALCULUS

MULTIVARIABLE

FIFTH EDITION

William G. McCallum
University of Arizona

Deborah Hughes-Hallett
University of Arizona

Andrew M. Gleason
Harvard University

et al.

Prepared by:

Rick Cangelosi
Scott Clark
Elliot J. Marks
Aaron Wootton

WILEY

John Wiley & Sons, Inc.

COVER PHOTO © Patrick Zephyr / Patrick Zephyr Nature Photography

To order books or for customer service please, call 1-800-CALL WILEY (225-5945).

This material is based upon work supported by the National Science Foundation under Grant No. DUE-9352905. Opinions expressed are those of the authors and not necessarily those of the Foundation.

ISBN-13   978-0-470-41413-2

Printed in the United States of America

10 9 8 7 6 5 4 3 2 1

Printed and bound by Bind-Rite Robbinsville

Table of Contents

CHAPTER TWELVE

Solutions for Section 12.1

Exercises

1. The distance of a point $P = (x, y, z)$ from the yz-plane is $|x|$, from the xz-plane is $|y|$, and from the xy-plane is $|z|$. So A is closest to the yz-plane, since it has the smallest x-coordinate in absolute value. B lies on the xz-plane, since its y-coordinate is 0. C is farthest from the xy-plane, since it has the largest z-coordinate in absolute value.

5. Your final position is $(1, -1, 1)$. This places you in front of the yz-plane, to the left of the xz-plane, and above the xy-plane.

9. The graph is all points with $y = 4$ and $z = 2$, i.e., a line parallel to the x-axis and passing through the points $(0, 4, 2); (2, 4, 2); (4, 4, 2)$ etc. See Figure 12.1.

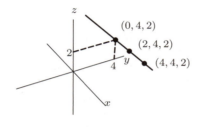

Figure 12.1

13. (a) 80-90°F
    (b) 60-72°F
    (c) 60-100°F

17. Table 12.1 gives the amount M spent on beef per household per week. Thus, the amount the household spent on beef in a year is $52M$. Since the household's annual income is I thousand dollars, the proportion of income spent on beef is

$$P = \frac{52M}{1000I} = 0.052\frac{M}{I}.$$

Thus, we need to take each entry in Table 12.1, divide it by the income at the left, and multiply by 0.052. Table 12.2 shows the results.

Table 12.1   *Money spent on beef ($/household/week)*

	Price of Beef ($)			
	3.00	3.50	4.00	4.50
20	7.95	9.07	10.04	10.94
40	12.42	14.18	15.76	17.46
60	15.33	17.50	19.88	21.78
80	16.05	18.52	20.76	22.82
100	17.37	20.20	22.40	24.89

Income ($1,000)

Table 12.2   *Proportion of annual income spent on beef*

	Price of Beef ($)			
	3.00	3.50	4.00	4.50
20	0.021	0.024	0.026	0.028
40	0.016	0.018	0.020	0.023
60	0.013	0.015	0.017	0.019
80	0.010	0.012	0.013	0.015
100	0.009	0.011	0.012	0.013

Income ($1,000)

Problems

21.

Table 12.3 *Temperature adjusted for wind-chill at 5 mph*

Temperature (°F)	35	30	25	20	15	10	5	0
Adjusted temperature (°F)	31	25	19	13	7	1	−5	−11

Table 12.4 *Temperature adjusted for wind-chill at 20 mph*

Temperature (°F)	35	30	25	20	15	10	5	0
Adjusted temperature (°F)	24	17	11	4	−2	−9	−15	−22

25. The gravitational force on a 100 kg object which is $7,000,000$ meters from the center of the earth (or about 600 km above the earth's surface) is about 820 newtons.

29. The distance of any point with coordinates (x, y, z) from the x-axis is $\sqrt{y^2 + z^2}$. The distance of the point from the xy-plane is $|x|$. Since the condition states that these distances are equal, the equation for the condition is

$$\sqrt{y^2 + z^2} = |x| \qquad \text{i.e.} \qquad y^2 + z^2 = x^2.$$

This is the equation of a cone whose tip is at the origin and which opens along the x-axis with a slope of 1 as shown in Figure 12.2.

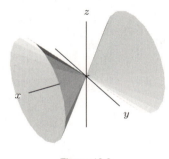

Figure 12.2

33. Using the distance formula, we find that

$$\text{Distance from } P_1 \text{ to } P = \sqrt{206}$$
$$\text{Distance from } P_2 \text{ to } P = \sqrt{152}$$
$$\text{Distance from } P_3 \text{ to } P = \sqrt{170}$$
$$\text{Distance from } P_4 \text{ to } P = \sqrt{113}$$

So $P_4 = (-4, 2, 7)$ is closest to $P = (6, 0, 4)$.

Solutions for Section 12.2

Exercises

1. (a) is (IV), since $z = 2 + x^2 + y^2$ is a paraboloid opening upward with a positive z-intercept.
(b) is (II), since $z = 2 - x^2 - y^2$ is a paraboloid opening downward.
(c) is (I), since $z = 2(x^2 + y^2)$ is a paraboloid opening upward and going through the origin.
(d) is (V), since $z = 2 + 2x - y$ is a slanted plane.
(e) is (III), since $z = 2$ is a horizontal plane.

5. The graph is a bowl opening up, with vertex at the point $(0, 0, 4)$. See Figure 12.3.

Figure 12.3   **Figure 12.4**

9. In the xy-plane, the graph is a circle of radius 2. Since there are no restrictions on z, we extend this circle along the z-axis. The graph is a circular cylinder extended in the z-direction. See Figure 12.4.

Problems

13. The one-variable function $f(a, t)$ represents the effect of an injection of a mg at time t. Figure 12.5 shows the graphs of the four functions $f(1, t) = te^{-4t}$, $f(2, t) = te^{-3t}$, $f(3, t) = te^{-2t}$, and $f(4, t) = te^{-t}$ corresponding to injections of 1, 2, 3, and 4 mg of the drug. The general shape of the graph is the same in every case: The concentration in the blood is zero at the time of injection $t = 0$, then increases to a maximum value, and then decreases toward zero again. We see that if a larger dose of the drug is administered, the peak of the graph is later and higher. This makes sense, since a larger dose will take longer to diffuse fully into the bloodstream and will produce a higher concentration when it does.

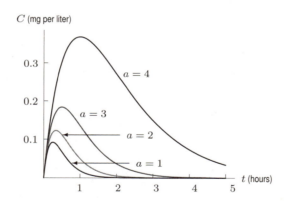

Figure 12.5: Concentration $C = f(a, t)$ of the drug resulting from an a mg injection

17. (a) This is a bowl; z increases as the distance from the origin increases, from a minimum of 0 at $x = y = 0$.
 (b) Neither. This is an upside-down bowl. This function will decrease from 1, at $x = y = 0$, to arbitrarily large negative values as x and y increase due to the negative squared terms of x and y. It will look like the bowl in part (a) except flipped over and raised up slightly.
 (c) This is a plate. Solving the equation for z gives $z = 1 - x - y$ which describes a plane whose x and y slopes are -1. It is perfectly flat, but not horizontal.
 (d) Within its domain, this function is a bowl. It is undefined at points at which $x^2 + y^2 > 5$, but within those limits it describes the bottom half of a sphere of radius $\sqrt{5}$ centered at the origin.
 (e) This function is a plate. It is perfectly flat and horizontal.

21. One possible equation: $z = (x - y)^2$. See Figure 12.6.

Figure 12.6

25. (a) If we have iron stomachs and can consume cola and pizza endlessly without ill effects, then we expect our happiness to increase without bound as we get more cola and pizza. Graph (IV) shows this since it increases along both the pizza and cola axes throughout.

(b) If we get sick upon eating too many pizzas or drinking too much cola, then we expect our happiness to decrease once either or both of those quantities grows past some optimum value. This is depicted in graph (I) which increases along both axes until a peak is reached, and then decreases along both axes.

(c) If we do get sick after too much cola, but are always able to eat more pizza, then we expect our happiness to decrease after we drink some optimum amount of cola, but continue to increase as we get more pizza. This is shown by graph (III) which increases continuously along the pizza axis but, after reaching a maximum, begins to decrease along the cola axis.

29. The paraboloid is $z = x^2 + y^2 + 5$, so it is represented by

$$z = f(x, y) = x^2 + y^2 + 5$$

and

$$g(x, y, z) = x^2 + y^2 + 5 - z = 0.$$

Other answers are possible.

Solutions for Section 12.3

Exercises

1. We'll set $z = 4$ at the peak. See Figure 12.7.

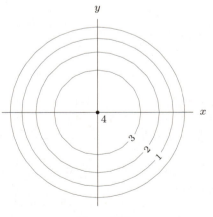

Figure 12.7

5. The contour where $f(x, y) = x + y = c$, or $y = -x + c$, is the graph of the straight line with slope -1 as shown in Figure 12.8. Note that we have plotted the contours for $c = -3, -2, -1, 0, 1, 2, 3$. The contours are evenly spaced.

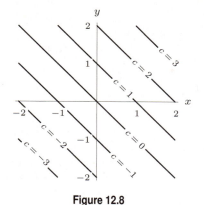

Figure 12.8

9. The contour where $f(x, y) = xy = c$, is the graph of the hyperbola $y = c/x$ if $c \neq 0$ and the coordinate axes if $c = 0$, as shown in Figure 12.9. Note that we have plotted contours for $c = -5, -4, -3, -2, -1, 0, 1, 2, 3, 4, 5$. The contours become more closely packed as we move further from the origin.

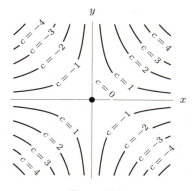

Figure 12.9

13. The contour where $f(x, y) = \cos(\sqrt{x^2 + y^2}) = c$, where $-1 \leq c \leq 1$, is a set of circles centered at $(0, 0)$, with radius $\cos^{-1} c + 2k\pi$ with $k = 0, 1, 2, ..$ and $-\cos^{-1} c + 2k\pi$, with $k = 1, 2, 3, ...$ as shown in Figure 12.10. Note that we have plotted contours for $c = 0, 0.2, 0.4, 0.6, 0.8, 1$.

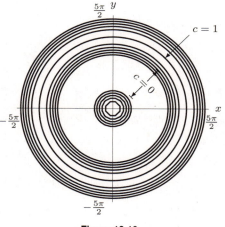

Figure 12.10

17. The values in Table 12.5 are not constant along rows or columns and therefore cannot be the lines shown in (I) or (IV). Also observe that as you move away from the origin, whose contour value is 0, the z-values on the contours increase. Thus, this table corresponds to diagram (II).

The values in Table 12.6 are also not constant along rows or columns. Since the contour values are decreasing as you move away from the origin, this table corresponds to diagram (III).

Table 12.7 shows that for each fixed value of x, we have constant contour value, suggesting a straight vertical line at each x-value, as in diagram (IV).

Table 12.8 also shows lines, however these are horizontal since for each fixed value of y we have constant contour values. Thus, this table matches diagram (I).

Problems

21. (a) The contour lines are much closer together on path A, so path A is steeper.

(b) If you are on path A and turn around to look at the countryside, you find hills to your left and right, obscuring the view. But the ground falls away on either side of path B, so you are likely to get a much better view of the countryside from path B.

(c) There is more likely to be a stream alongside path A, because water follows the direction of steepest descent.

25. Figure 12.11 shows an east-west cross-section along the line $N = 50$ kilometers.

Figure 12.12 shows an east-west cross-section along the line $N = 100$ kilometers.

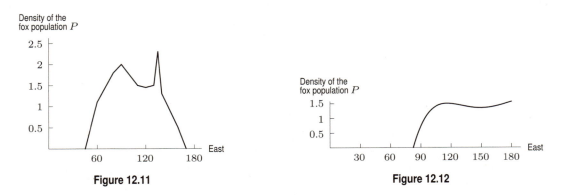

Figure 12.11   **Figure 12.12**

Figure 12.13 shows a north-south cross-section along the line $E = 60$ kilometers.

Figure 12.14 shows a north-south cross-section along the line $E = 120$ kilometers.

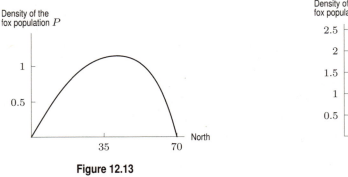   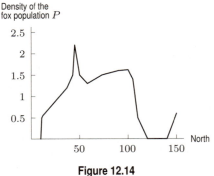

Figure 12.13   **Figure 12.14**

29. (a) Multiply the values on each contour of the original contour diagram by 3. See Figure 12.15.

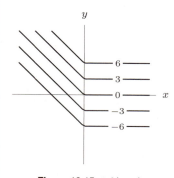

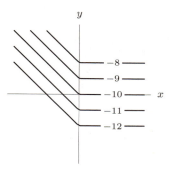

Figure 12.15: $3f(x, y)$

Figure 12.16: $f(x, y) - 10$

(b) Subtract 10 from the values on each contour. See Figure 12.16.
(c) Shift the diagram 2 units to the right and 2 units up. See Figure 12.17.

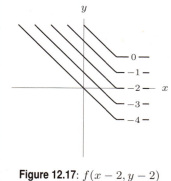

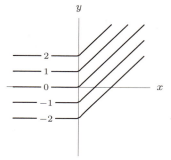

Figure 12.17: $f(x - 2, y - 2)$

Figure 12.18: $f(-x, y)$

(d) Reflect the diagram about the y-axis. See Figure 12.18.

33. Since $f(x, y) = x^2 - y^2 = (x - y)(x + y) = 0$ gives $x - y = 0$ or $x + y = 0$, the contours $f(x, y) = 0$ are the lines $y = x$ or $y = -x$. In the regions between them, $f(x, y) > 0$ or $f(x, y) < 0$ as shown in Figure 12.19. The surface $z = f(x, y)$ is above the xy-plane where $f > 0$ (that is on the shaded regions containing the x-axis) and is below the xy-plane where $f < 0$. This means that a person could sit on the surface facing along the positive or negative x-axis, and with his/her legs hanging down the sides below the y-axis. Thus, the graph of the function is saddle-shaped at the origin.

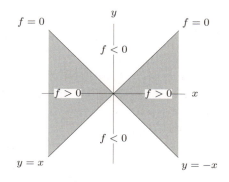

Figure 12.19

Solutions for Section 12.4

Exercises

1.

Table 12.5

$x\backslash y$	0.0	1.0
0.0	−1.0	1.0
2.0	3.0	5.0

5. A table of values is linear if the rows are all linear and have the same slope and the columns are all linear and have the same slope. The table might represent a linear function since the slope in each row is 5 and the slope in each column is 2.

9. Since

$$0 = c + m \cdot 0 + n \cdot 0 \qquad c = 0$$
$$-1 = c + m \cdot 0 + n \cdot 2 \qquad c + 2n = -1$$
$$-4 = c + m \cdot (-3) + n \cdot 0 \quad c - 3m = -4$$

we get:

$$c = 0, m = \frac{4}{3}, n = -\frac{1}{2}.$$

Thus, $z = \frac{4}{3}x - \frac{1}{2}y$.

13. (a) Since z is a linear function of x and y with slope 2 in the x-direction, and slope 3 in the y-direction, we have:

$$z = 2x + 3y + c$$

We can write an equation for changes in z in terms of changes in x and y:

$$\Delta z = (2(x + \Delta x) + 3(y + \Delta y) + c) - (2x + 3y + c)$$
$$= 2\Delta x + 3\Delta y$$

Since $\Delta x = 0.5$ and $\Delta y = -0.2$, we have

$$\Delta z = 2(0.5) + 3(-0.2) = 0.4$$

So a 0.5 change in x and a -0.2 change in y produces a 0.4 change in z.

(b) As we know that $z = 2$ when $x = 5$ and $y = 7$, the value of z when $x = 4.9$ and $y = 7.2$ will be

$$z = 2 + \Delta z = 2 + 2\Delta x + 3\Delta y$$

where Δz is the change in z when x changes from 4.9 to 5 and y changes from 7.2 to 7. We have $\Delta x = 4.9 - 5 = -0.1$ and $\Delta y = 7.2 - 7 = 0.2$. Therefore, when $x = 4.9$ and $y = 7.2$, we have

$$z = 2 + 2 \cdot (-0.1) + 5 \cdot 0.2 = 2.4$$

Problems

17. (a) Expenditure, E, is given by the equation:

$$E = (\text{price of raw material 1})m_1 + (\text{price of raw material 2})m_2 + C$$

where C denotes all the other expenses (assumed to be constant). Since the prices of the raw materials are constant, but m_1 and m_2 are variables, we have a linear function.

(b) Revenue, R, is given by the equation:

$$R = (p_1)q_1 + (p_2)q_2.$$

Since p_1 and p_2 are constant, while q_1 and q_2 are variables, we again have a linear function.

(c) Revenue is again given by the equation,

$$R = (p_1)q_1 + (p_2)q_2.$$

Since p_2 and q_2 are now constant, the term $(p_2)q_2$ is also constant. However, since p_1 and q_1 are variables, the $(p_1)q_1$ term means that the function is not linear.

21. The function, g, has a slope of 3 in the x direction and a slope of 1 in the y direction, so $g(x, y) = c + 3x + y$. Since $g(0, 0) = 0$, the formula is $g(x, y) = 3x + y$.

25. See Figure 12.20.

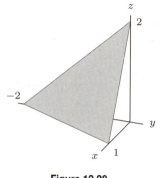

Figure 12.20

29. (a) The contours of f have equation

$$k = c + mx + ny, \quad \text{where } k \text{ is a constant.}$$

Solving for y gives:

$$y = -\frac{m}{n}x + \frac{k - c}{n}$$

Since c, m, n and k are constants, this is the equation of a line. The coefficient of x is the slope and is equal to $-m/n$.

(b) Substituting $x + n$ for x and $y - m$ for y into $f(x, y)$ gives

$$f(x + n, y - m) = c + m(x + n) + n(y - m)$$

Multiplying out and simplifying gives

$$f(x + n, y - m) = c + mx + mn + ny - nm$$

$$f(x + n, y - m) = c + mx + ny = f(x, y)$$

(c) Part (b) tells us that if we move n units in the x direction and $-m$ units in the y direction, the value of the function $f(x, y)$ remains constant. Since contours are lines where the function has a constant value, this implies that we remain on the same contour. This agrees with part (a) which tells us that the slope of any contour line will be $-m/n$. Since the slope is $\Delta y/\Delta x$, it follows that changing y by $-m$ and x by n will keep us on the same contour.

Solutions for Section 12.5

Exercises

1. (a) Observe that setting $f(x, y, z) = c$ gives a cylinder about the x-axis, with radius $\sqrt{c}$. These surfaces are in graph (I).

(b) By the same reasoning the level curves for $h(x, y, z)$ are cylinders about the y-axis, so they are represented in graph (II).

5. If we solve for z, we get $z = (1 - x^2 - y)^2$, so the level surface is the graph of $f(x, y) = (1 - x^2 - y)^2$.

9. A hyperboloid of two sheets.

13. Yes,

$$z = f(x, y) = \frac{2}{5}x + \frac{3}{5}y - 2.$$

Problems

17. The top half of the sphere is represented by

$$z = f(x, y) = \sqrt{10 - x^2 - y^2}$$

and

$$g(x, y, z) = x^2 + y^2 + z^2 = 10, \quad z \geq 0.$$

Other answers are possible.

21. The equation of any plane parallel to the plane $z = 2x+3y-5$ has x-slope 2 and y-slope 3, so has equation $z = 2x+3y-c$ for any constant c, or $2x + 3y - z = c$. Thus we could take $g(x, y, z) = 2x + 3y - z$. Other answers are possible.

25. In the xz-plane, the equation $x^2/4 + z^2 = 1$ is an ellipse, with widest points at $x = \pm 2$ on the x-axis and crossing the z-axis at $z = \pm 1$. Since the equation has no y term, the level surface is a cylinder of elliptical cross-section, centered along the y-axis.

29. Let's consider the function $y = 2 + \sin z$ drawn in the yz-plane in Figure 12.21.

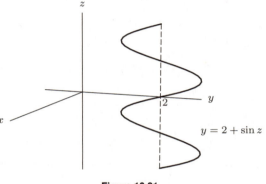

Figure 12.21

Now rotate this graph around the z-axis. Then, a point (x, y, z) is on the surface if and only if $x^2 + y^2 = (2 + \sin z)^2$. Thus, the surface generated is a surface of rotation with the profile shown in Figure 12.21.

Similarly, the surface with equation $x^2 + y^2 = (f(z))^2$ is the surface obtained rotating the graph of $y = f(z)$ around the z-axis.

33. For values of $f < 4$, the level surfaces are spheres, with larger f giving smaller radii. See Figure 12.22.

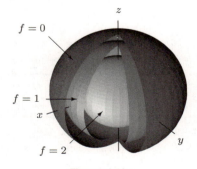

Figure 12.22

Solutions for Section 12.6

Exercises

1. No, $1/(x^2 + y^2)$ is not defined at the origin, so is not continuous at all points in the square $-1 \le x \le 1, -1 \le y \le 1$.

5. The function $\tan(\theta)$ is undefined when $\theta = \pi/2 \approx 1.57$. Since there are points in the square $-2 \le x \le 2, -2 \le y \le 2$ with $x \cdot y = \pi/2$ (e.g. $x = 1, y = \pi/2$) the function $\tan(xy)$ is not defined inside the square, hence not continuous.

9. Since f does not depend on y we have:

$$\lim_{(x,y)\to(0,0)} f(x,y) = \lim_{x\to 0} \frac{x}{x^2 + 1} = \frac{0}{0 + 1} = 0.$$

Problems

13. We want to show that f does not have a limit as (x, y) approaches $(0, 0)$. Let us suppose that (x, y) tends to $(0, 0)$ along the line $y = mx$. Then

$$f(x, y) = f(x, mx) = \frac{x^2 - m^2 x^2}{x^2 + m^2 x^2} = \frac{1 - m^2}{1 + m^2}.$$

Therefore

$$\lim_{x\to 0} f(x, mx) = \frac{1 - m^2}{1 + m^2}$$

and so for $m = 1$ we get

$$\lim_{\substack{(x,y)\to(0,0)\\y=x}} f(x, y) = \frac{1 - 1}{1 + 1} = \frac{0}{2} = 0$$

and for $m = 0$

$$\lim_{\substack{(x,y)\to(0,0)\\y=0}} f(x, y) = \frac{1 - 0}{1 + 0} = 1.$$

Thus no matter how close they are to the origin, there will be points (x, y) such that $f(x, y)$ is close to 0 and points (x, y) where $f(x, y)$ is close to 1. So the limit:

$$\lim_{(x,y)\to(0,0)} f(x, y) \text{ does not exist.}$$

17. We will study the continuity of f at $(a, 0)$. Now $f(a, 0) = 1 - a$. In addition:

$$\lim_{\substack{(x,y)\to(a,0)\\y>0}} f(x, y) = \lim_{x\to a}(1 - x) = 1 - a$$

$$\lim_{\substack{(x,y)\to(a,0)\\y<0}} f(x, y) = \lim_{x\to a} -2 = -2.$$

If $a = 3$, then

$$\lim_{\substack{(x,y)\to(3,0)\\y>0}} f(x, y) = 1 - 3 = -2 = \lim_{\substack{(x,y)\to(3,0)\\y<0}} f(x, y)$$

and so $\lim_{(x,y)\to(3,0)} f(x, y) = -2 = f(3, 0)$. Therefore f is continuous at $(3, 0)$.

On the other hand, if $a \ne 3$, then

$$\lim_{\substack{(x,y)\to(a,0)\\y>0}} f(x, y) = 1 - a \ne -2 = \lim_{\substack{(x,y)\to(a,0)\\y<0}} f(x, y)$$

so $\lim_{(x,y)\to(a,0)} f(x, y)$ does not exist. Thus f is not continuous at $(a, 0)$ if $a \ne 3$.

Thus, f is not continuous along the line $y = 0$. (In fact the only point on this line where f is continuous is the point $(3, 0)$.)

21. The function $f(x, y) = x^2 + y^2 + 1$ gets closer and closer to 1 as (x, y) gets closer to the origin. To make f continuous at the origin, we need to have $f(0, 0) = 1$. Thus $c = 1$ will make the function continuous at the origin.

Solutions for Chapter 12 Review

Exercises

1. An example is the line $z = -x$ in the xz-plane. See Figure 12.23.

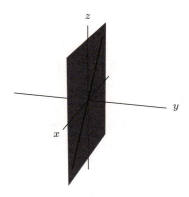

Figure 12.23

5. Given (x, y) we can solve uniquely for z, namely $z = 2 + \dfrac{x}{5} + \dfrac{y}{5} - \dfrac{3x^2}{5} + y^2$. Thus, z is a function of x and y:

$$z = f(x, y) = 2 + \frac{x}{5} + \frac{y}{5} - \frac{3x^2}{5} + y^2.$$

9. Contours are lines of the form $3x - 5y + 1 = c$ as shown in Figure 12.24. Note that for the regions of x and y given, the c values range from $-12 < c < 12$ and are evenly spaced.

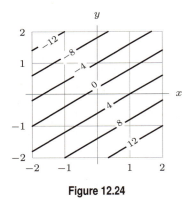

Figure 12.24

13. These conditions describe a line parallel to the z-axis which passes through the xy-plane at $(2, 1, 0)$.

17. A contour diagram is linear if the contours are parallel straight lines, equally spaced for equally spaced values of z. This contour diagram does not represent a linear function.

21. The level surfaces appear to be circular cylinders centered on the z-axis. Since they don't change with z, there is no z in the formula, and we can use the formula for a circle in the xy-plane, $x^2 + y^2 = r^2$. Thus the level surfaces are of the form $f(x, y, z) = x^2 + y^2 = c$ for $c > 0$.

25. (a) The value of z decreases as x increases. See Figure 12.25.
   (b) The value of z increases as y increases. See Figure 12.26.

Figure 12.25   **Figure 12.26**

Problems

29. Points along the positive x-axis are of the form $(x, 0)$; at these points the function looks like $2x/2x = 1$ everywhere (except at the origin, where it is undefined). On the other hand, along the y-axis, the function looks like $-y^2/y^2 = -1$. Since approaching the origin along two different paths yields numbers that are not the same, the limit does not exist.

33. When h is fixed, say $h = 1$, then

$$V = f(r, 1) = \pi r^2 1 = \pi r^2$$

Similarly,

$$f(r, \frac{2}{3}) = \frac{4}{9}\pi r^2 \quad \text{and} \quad f(r, \frac{1}{3}) = \frac{\pi}{9}r^2$$

When r is fixed, say $r = 1$, then

$$f(1, h) = \pi(1)^2 h = \pi h$$

Similarly,

$$f(2, h) = 4\pi \quad \text{and} \quad f(3, h) = 9\pi h.$$

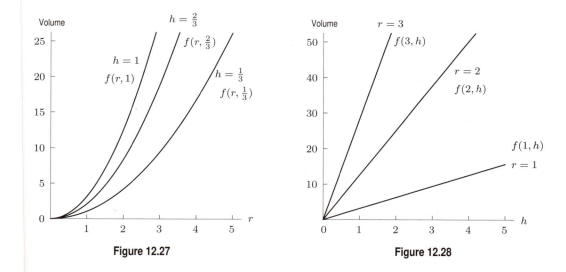

Figure 12.27   **Figure 12.28**

37. The function $f(0, t) = \cos t \sin 0 = 0$ gives the displacement of the left end of the string as time varies. Since that point remains stationary, the displacement is zero. The function $f(1, t) = \cos t \sin 1 = 0.84 \cos t$ gives the displacement of the point at $x = 1$ as time varies. Since $\cos t$ oscillates back and forth between 1 and -1, this point moves back and forth with maximum displacement of 0.84 in either direction. Notice the maximum displacements are greatest at $x = \pi/2$ where $\sin x = 1$.

CAS Challenge Problems

41. (a)

$$f(x, f(x, y)) = 3 + x + 2(3 + x + 2y) = (3 + 2 \cdot 3) + (1 + 2)x + 2^2 y = 9 + 3x + 4y$$
$$f(x, f(x, f(x, y))) = 3 + x + 2(3 + x + 2(3 + x + 2y))$$
$$= (3 + 2 \cdot 3 + 2^2 \cdot 3) + (1 + 2 + 2^2)x + 2^3 y = 21 + 7x + 8y$$

(b) From part (a) we guess that the general pattern for k nested fs is

$$(3 + 2 \cdot 3 + 2^2 \cdot 3 + \cdots + 2^{k-1} \cdot 3) + (1 + 2 + 2^2 + \cdots + 2^{k-1})x + 2^k y$$

Thus

$$f(x, f(x, f(x, f(x, f(x, f(x, y)))))) =$$
$$(3 + 2 \cdot 3 + 2^2 \cdot 3 + \cdots + 2^5 \cdot 3) + (1 + 2 + 2^2 + \cdots + 2^5)x + 2^6 y = 189 + 63x + 64y.$$

CHECK YOUR UNDERSTANDING

1. Could not be true. If the origin is on the level curve $z = 1$, then $z = f(0,0) = 1 \neq -1$. So $(0,0)$ cannot be on both $z = 1$ and $z = -1$.

5. True. For every point (x, y), compute the value $z = e^{-(x^2 + y^2)}$ at that point. The level curve obtained by getting z equal to that value goes through the point (x, y).

9. False. If, for example, $d = 2$ meters and $H = 57°C$, there could be many times t at which the water temperature is $57\,°C$ at 2 meters depth.

13. True. If there were such an intersection point, that point would have two different temperatures simultaneously.

17. False. The point $(0, 0, 0)$ does not satisfy the equation.

21. True. The x-axis is where $y = z = 0$.

25. False. If $x = 10$, substituting gives $10^2 + y^2 + z^2 = 10$, so $y^2 + z^2 = -90$. Since $y^2 + z^2$ cannot be negative, a point with $x = 10$ cannot satisfy the equation.

29. True. The cross-sections with $y = c$ are of the form $z = 1 - c^2$, which are horizontal lines.

33. False. Wherever $f(x, y) = 0$ the graphs of $f(x, y)$ and $-f(x, y)$ will intersect.

37. True. If $f = c$ then the contours are of the form $c = y^2 + (x - 2)^2$, which are circles centered at $(2, 0)$ if $c > 0$. But if $c = 0$ the contour is the single point $(2, 0)$.

41. False. As a counterexample, consider any function with one variable missing, e.g. $f(x, y) = x^2$. The graph of this is not a plane (it is a *parabolic cylinder*) but has contours which are lines of the form $x = c$.

45. True. The graph of g is the same as the graph of f translated down by 5 units, so the horizontal slice of f at height 5 is the same as the horizontal slice of g at height 0.

49. True. $f(0, 0) = 0$, $f(0, 1) = 4$ give a y slope of 4, but $f(0, 0) = 0$, $f(0, 3) = 5$ give a y slope of 5/3. Since linearity means the y slope must be the same between any two points, this function cannot be linear.

53. True. Functions can have only one value for a given input, so their graphs can intersect a vertical line at most once. A vertical plane would not satisfy this property, so cannot be the graph of a function.

57. True. Both are the set of all points (x, y, z) in 3-space satisfying $z = x^2 + y^2$.

61. True. The level surfaces are of the form $x + 2y + z = k$, or $z = k - x - 2y$. These are the graphs of the linear functions $f(x, y) = k - x - 2y$, each of which has x-slope of -1 and y-slope equal to -2. Thus they form parallel planes.

65. False. For example, the function $g(x, y, z) = \sin(x + y + z)$ has level surfaces of the form $x + y + z = k$, where $k = \arcsin(c) + n\pi$, for $n = 0, \pm1, \pm2, \ldots$. These surfaces are planes (for $-1 \leq c \leq 1$).

CHAPTER THIRTEEN

Solutions for Section 13.1

Exercises

1. The vectors are $\vec{a} = \vec{i} + 3\vec{j}$, $\vec{b} = 3\vec{i} + 2\vec{j}$, $\vec{v} = -2\vec{i} - 2\vec{j}$, and $\vec{w} = -\vec{i} + 2\vec{j}$.

5. $\vec{a} = \vec{b} = \vec{c} = 3\vec{k}$, $\vec{d} = 2\vec{i} + 3\vec{k}$, $\vec{e} = \vec{j}$, $\vec{f} = -2\vec{i}$

9. $-4\vec{i} + 8\vec{j} - 0.5\vec{i} + 0.5\vec{k} = -4.5\vec{i} + 8\vec{j} + 0.5\vec{k}$

13. $0.6\vec{i} + 0.2\vec{j} - \vec{k} + 0.3\vec{i} + 0.3\vec{k} = 0.9\vec{i} + 0.2\vec{j} - 0.7\vec{k}$

17. $\|\vec{v}\| = \sqrt{1^2 + (-1)^2 + 3^2} = \sqrt{11}$.

21.

$$
\begin{aligned}
5\vec{a} + 2\vec{b} &= 5(2\vec{j} + \vec{k}) + 2(-3\vec{i} + 5\vec{j} + 4\vec{k}) \\
&= (5(2)\vec{j} + 5(1)\vec{k}) + (2(-3)\vec{i} + 2(5)\vec{j} + 2(4)\vec{k}) \\
&= (10\vec{j} + 5\vec{k}) + (-6\vec{i} + 10\vec{j} + 8\vec{k}) = (0-6)\vec{i} + (10+10)\vec{j} + (5+8)\vec{k} \\
&= -6\vec{i} + 20\vec{j} + 13\vec{k}.
\end{aligned}
$$

25.

$$
\begin{aligned}
\|\vec{y} - \vec{x}\| &= \|(4\vec{i} - 7\vec{j}) - (-2\vec{i} + 9\vec{j})\| = \|(4-(-2))\vec{i} + (-7-9)\vec{j}\| = \|6\vec{i} - 16\vec{j}\| \\
&= \sqrt{6^2 + (-16)^2} = \sqrt{36 + 256} = \sqrt{292} = 2\sqrt{73}.
\end{aligned}
$$

Problems

29. If two vectors are parallel, they are scalar multiples of one another. Thus

$$
\frac{a^2}{5a} = \frac{6}{-3}.
$$

Solving for a gives

$$
a^2 = -2 \cdot 5a \quad \text{so} \quad a = 0, -10.
$$

33. (a) The components are $v_1 = 2\cos\pi/4 = \sqrt{2}$, $v_2 = 2\sin\pi/4 = \sqrt{2}$. See Figure 13.1. Thus $\vec{v} = \sqrt{2}\vec{i} + \sqrt{2}\vec{j}$.

(b) Since the vector lies in the xz-plane, its y-component is 0. Its x-component is $1\cos(\frac{\pi}{6})\vec{i} = \frac{\sqrt{3}}{2}\vec{i}$ and its z-component is $1\sin(\frac{\pi}{6})\vec{j} = \frac{1}{2}\vec{j}$. See Figure 13.2. So the vector is $\frac{\sqrt{3}}{2}\vec{i} + \frac{1}{2}\vec{k}$.

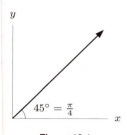

Figure 13.1

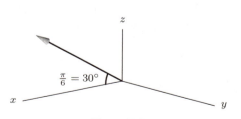

Figure 13.2

37.

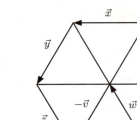

Figure 13.3

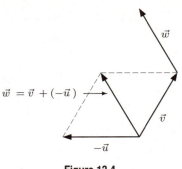

$$\vec{w} = \vec{v} + (-\vec{u})$$

Figure 13.4

Break the hexagon up into 6 equilateral triangles, as shown in Figure 13.3.
Then $\vec{u} - \vec{v} + \vec{w} = \vec{0}$, so $\vec{w} = \vec{v} - \vec{u}$
Similarly, $\vec{x} = -\vec{u}$, $\vec{y} = -\vec{v}$, $\vec{z} = -\vec{w} = \vec{u} - \vec{v}$.

41.

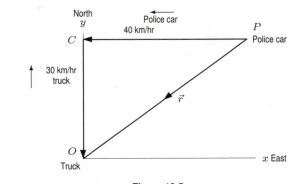

Figure 13.5

Since both vehicles reach the crossroad in exactly one hour, at the present the truck is at O in Figure 13.5; the police car is at P and the crossroads is at C. If $\vec{r}$ is the vector representing the line of sight of the truck with respect to the police car.

$$\vec{r} = -40\vec{i} - 30\vec{j}$$

Solutions for Section 13.2

Exercises

1. Scalar

5. Writing $\vec{P} = (P_1, P_2, \cdots, P_{50})$ where P_i is the population of the i-th state, shows that $\vec{P}$ can be thought of as a vector with 50 components.

9. We need to calculate the length of each vector.

$$\|21\vec{i} + 35\vec{j}\| = \sqrt{21^2 + 35^2} = \sqrt{1666} \approx 40.8,$$
$$\|40\vec{i}\| = \sqrt{40^2} = 40.$$

So the first car is faster.

Problems

13. The velocity vector of the plane with respect to the air has the form

$$\vec{v} = a\vec{i} + 80\vec{k} \quad \text{where } \|\vec{v}\| = 480.$$

(See Figure 13.6.) Therefore $\sqrt{a^2 + 80^2} = 480$ so $a = \sqrt{480^2 - 80^2} \approx 473.3$ km/hr. We conclude that $\vec{v} \approx 473.3\vec{i} + 80\vec{k}$.

The wind vector is

$$\vec{w} = 100(\cos 45°)\vec{i} + 100(\sin 45°)\vec{j}$$
$$\approx 70.7\vec{i} + 70.7\vec{j}$$

The velocity vector of the plane with respect to the ground is then

$$\vec{v} + \vec{w} = (473.3\vec{i} + 80\vec{k}) + (70.7\vec{i} + 70.7\vec{j})$$
$$= 544\vec{i} + 70.7\vec{j} + 80\vec{k}$$

From Figure 13.7, we see that the velocity relative to the ground is

$$544\vec{i} + 70.7\vec{j}.$$

The ground speed is therefore $\sqrt{544^2 + 70.7^2} \approx 548.6$ km/hr.

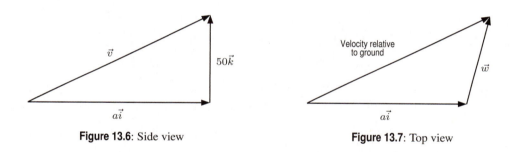

Figure 13.6: Side view   **Figure 13.7**: Top view

17. Let $\vec{R}$ be the resultant force, and let $\vec{F}_1$ and $\vec{F}_2$ be the forces exerted by the larger and smaller tugs. See Figure 13.8. Then $\|\vec{F}_1\| = \frac{5}{4}\|\vec{F}_2\|$. The y components of the vectors $\vec{F}_1$ and $\vec{F}_2$ must cancel each other in order to ensure that the ship travels due east, hence

$$\|\vec{F}_1\| \sin 30° = \|\vec{F}_2\| \sin \theta,$$

so

$$\frac{5}{4}\|\vec{F}_2\| \sin 30° = \|\vec{F}_2\| \sin \theta,$$

giving $\sin \theta = \frac{5}{8}$, and hence $\theta = \sin^{-1}\frac{5}{8} = 38.7°$.

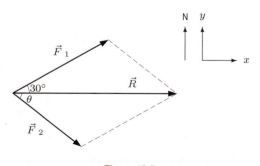

Figure 13.8

21. We want the total force on the object to be zero. We must choose the third force $\vec{F}_3$ so that $\vec{F}_1 + \vec{F}_2 + \vec{F}_3 = 0$. Since $\vec{F}_1 + \vec{F}_2 = 11\vec{i} - 4\vec{j}$, we need $\vec{F}_3 = -11\vec{i} + 4\vec{j}$.

25. (a) Let x-axis be the East direction and y-axis be the North direction. From Figure 13.9,

$$\theta = \sin^{-1}(4/5) = 53.1°.$$

That is, he should steer at $53.1°$ east of south.

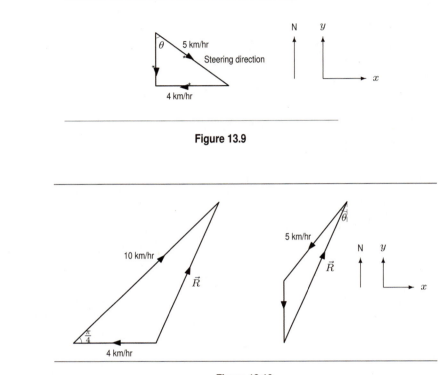

Figure 13.9

(b)

Figure 13.10

Let $\vec{R}$ be the resultant of the wind and river velocities, that is

$$\vec{R} = -4\vec{i} + (10\cos(\frac{\pi}{4})\vec{i} + 10\cos(\frac{\pi}{4})\vec{j})$$
$$= (-4 + 5\sqrt{2})\vec{i} + 5\sqrt{2}\vec{j}.$$

From Figure 13.10, we see that to get the the x-component of his rowing velocity and the x-component of $\vec{R}$ to cancel each other, we must have

$$5\sin\theta = -4 + 5\sqrt{2}$$
$$\theta = \sin^{-1}\left(\frac{-4 + 5\sqrt{2}}{5}\right) = 37.9°.$$

However for this value of θ, the y-component of the velocity is

$$5\sqrt{2} - 5\cos(37.9°) = 3.1.$$

Since the y-component is positive, the man will not move across the river in a southward direction.

29.

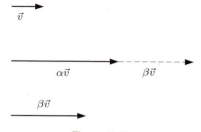

Figure 13.11

The vectors $\vec{v}$, $\alpha\vec{v}$ and $\beta\vec{v}$ are all parallel. Figure 13.11 shows them with $\alpha, \beta > 0$, so all the vectors are in the same direction. Notice that $\alpha\vec{v}$ is a vector α times as long as $\vec{v}$ and $\beta\vec{v}$ is β times as long as $\vec{v}$. Therefore $\alpha\vec{v} + \beta\vec{v}$ is a vector $(\alpha + \beta)$ times as long as $\vec{v}$, and in the same direction. Thus,

$$\alpha\vec{v} + \beta\vec{v} = (\alpha + \beta)\vec{v}.$$

33. According to the definition of scalar multiplication, $1 \cdot \vec{v}$ has the same direction and magnitude as $\vec{v}$, so it is the same as $\vec{v}$.

Solutions for Section 13.3

Exercises

1. $\vec{a} \cdot \vec{y} = (2\vec{j} + \vec{k}) \cdot (4\vec{i} - 7\vec{j}) = -14.$

5. $\vec{c} \cdot \vec{a} + \vec{a} \cdot \vec{y} = (\vec{i} + 6\vec{j}) \cdot (2\vec{j} + \vec{k}) + (2\vec{j} + \vec{k}) \cdot (4\vec{i} - 7\vec{j}) = 12 - 14 = -2.$

9. Since $\vec{c} \cdot \vec{c}$ is a scalar and $(\vec{c} \cdot \vec{c})\vec{a}$ is a vector, the answer to this equation is another scalar. We could calculate $\vec{c} \cdot \vec{c}$, then $(\vec{c} \cdot \vec{c})\vec{a}$, and then take the dot product $((\vec{c} \cdot \vec{c})\vec{a}) \cdot \vec{a}$. Alternatively, we can use the fact that

$$((\vec{c} \cdot \vec{c})\vec{a}) \cdot \vec{a} = (\vec{c} \cdot \vec{c})(\vec{a} \cdot \vec{a}).$$

Since

$$\vec{c} \cdot \vec{c} = (\vec{i} + 6\vec{j}) \cdot (\vec{i} + 6\vec{j}) = 1^2 + 6^2 = 37$$
$$\vec{a} \cdot \vec{a} = (2\vec{j} + \vec{k}) \cdot (2\vec{j} + \vec{k}) = 2^2 + 1^2 = 5,$$

we have,

$$(\vec{c} \cdot \vec{c})(\vec{a} \cdot \vec{a}) = 37(5) = 185$$

13. The equation can be rewritten as

$$z - 5x + 10 = 15 - 3y$$
$$-5x + 3y + z = 5$$

so $\vec{n} = -5\vec{i} + 3\vec{j} + \vec{k}.$

17. (a) Writing the plane in the form $2x + 3y - z = 0$ shows that a normal vector is

$$\vec{n} = 2\vec{i} + 3\vec{j} - \vec{k}.$$

Any multiple of this vector is also a correct answer.

(b) Any vector perpendicular to $\vec{n}$ is parallel to the plane, so one possible answer is

$$\vec{v} = 3\vec{i} - 2\vec{j}.$$

Many other answers are possible.

21. (a) We first find the unit vector in direction $\vec{v}$. Since $||\vec{v}|| = \sqrt{3^2 + 4^2} = 5$, the unit vector in direction of $\vec{v}$ is $\vec{u} = 0.6\vec{i} + 0.8\vec{j}$. Then

$$
\begin{aligned}
\vec{F}_{\text{parallel}} &= (\vec{F} \cdot \vec{u})\vec{u} \\
&= (-0.4 \cdot 0.6 + 0.3 \cdot 0.8)\vec{u} \\
&= \vec{0}.
\end{aligned}
$$

Notice that the component of $\vec{F}$ in direction $\vec{v}$ is equal to $\vec{0}$. This makes sense (and could have been predicted) since $\vec{F}$ is perpendicular to $\vec{v}$.

(b) We have
$$
\vec{F}_{\text{perp}} = \vec{F} - \vec{F}_{\text{parallel}} = \vec{F}.
$$

(c) Since work is the dot product of the force and displacement vectors, we have

$$
W = \vec{F} \cdot \vec{v} = -0.4 \cdot 3 + 0.3 \cdot 4 = 0.
$$

Notice that since the force is perpendicular to the displacement, the work done is zero.

25. (a) We first find the unit vector in direction $\vec{v}$. Since $||\vec{v}|| = \sqrt{5^2 + (-1)^2} = \sqrt{26}$, the unit vector in direction of $\vec{v}$ is $\vec{u} = \vec{v}/\sqrt{26}$. Then

$$
\begin{aligned}
\vec{F}_{\text{parallel}} &= (\vec{F} \cdot \vec{u})\vec{u} \\
&= (20/\sqrt{26})\vec{u} \\
&= \frac{100}{26}\vec{i} + \frac{-20}{26}\vec{j} \\
&= 3.846\vec{i} - 0.769\vec{j}.
\end{aligned}
$$

(b) We have
$$
\vec{F}_{\text{perp}} = \vec{F} - \vec{F}_{\text{parallel}} = (-20\vec{j}) - (3.846\vec{i} - 0.769\vec{j}) = -3.846\vec{i} - 19.231\vec{j}.
$$

(c) Since work is the dot product of the force and displacement vectors, we have

$$
W = \vec{F} \cdot \vec{v} = 20.
$$

Problems

29. (a) Perpendicular vectors have a dot product of 0. Since $\vec{a} \cdot \vec{c} = 1(-2) - 3(-1) - 1 \cdot 1 = 0$, and $\vec{b} \cdot \vec{d} = 1(-1) + 1(-1) + 2 \cdot 1 = 0$, the pairs we want are $\vec{a}, \vec{c}$ and $\vec{b}, \vec{d}$.
(b) Parallel vectors are multiples of one another, so there are no parallel vectors in this set.
(c) Since $\vec{v} \cdot \vec{w} = ||\vec{v}||||\vec{w}|| \cos\theta$, the dot product of the vectors we want is positive. We have

$$
\begin{aligned}
\vec{a} \cdot \vec{b} &= 1 \cdot 1 - 3 \cdot 1 - 1 \cdot 2 = -4 \\
\vec{a} \cdot \vec{d} &= 1(-1) - 3(-1) - 1 \cdot 1 = 1 \\
\vec{b} \cdot \vec{c} &= 1(-2) + 3(-1) + 2 \cdot 1 = -1 \\
\vec{c} \cdot \vec{d} &= -2(-1) - 1(-1) + 1 \cdot 1 = 4,
\end{aligned}
$$

and we already know $\vec{a} \cdot \vec{c} = \vec{b} \cdot \vec{d} = 0$. Thus, the pairs of vectors with an angle of less than $\pi/2$ between them are $\vec{a}, \vec{d}$ and $\vec{c}, \vec{d}$.
(d) Vectors with an angle of more than $\pi/2$ between them have a negative dot product, so pairs are $\vec{a}, \vec{b}$ and $\vec{b}, \vec{c}$.

33. If the vectors are perpendicular, we need

$$
\vec{v} \cdot \vec{w} = (2a\vec{i} - a\vec{j} + 16\vec{k}) \cdot (5\vec{i} + a\vec{j} - \vec{k}) = 10a - a^2 - 16 = 0.
$$

Solving $10a - a^2 - 16 = -(a - 2)(a - 8) = 0$ gives $a = 2, 8$.

37. The plane is

$$
\begin{aligned}
3x - y + 4z &= 3 \cdot 1 - 1 \cdot 5 + 4 \cdot 2 \\
3x - y + 4z &= 6.
\end{aligned}
$$

41. Two planes are parallel if their normal vectors are parallel. Since the plane $3x + y + z = 4$ has normal vector $\vec{n} = 3\vec{i} + \vec{j} + \vec{k}$, the plane we are looking for has the same normal vector and passes through the point $(-2, 3, 2)$. Thus, it has the equation

$$3x + y + z = 3 \cdot (-2) + 3 + 2 = -1.$$

45. (a) The points A, B and C are shown in Figure 13.12.

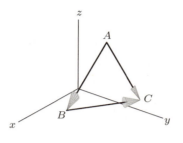

Figure 13.12

First, we calculate the vectors which form the sides of this triangle:
$\overrightarrow{AB} = (4\vec{i} + 2\vec{j} + \vec{k}) - (2\vec{i} + 2\vec{j} + 2\vec{k}) = 2\vec{i} - \vec{k}$
$\overrightarrow{BC} = (2\vec{i} + 3\vec{j} + \vec{k}) - (4\vec{i} + 2\vec{j} + \vec{k}) = -2\vec{i} + \vec{j}$
$\overrightarrow{AC} = (2\vec{i} + 3\vec{j} + \vec{k}) - (2\vec{i} + 2\vec{j} + 2\vec{k}) = \vec{j} - \vec{k}$

Now we calculate the lengths of each of the sides of the triangles:
$\|\overrightarrow{AB}\| = \sqrt{2^2 + (-1)^2} = \sqrt{5}$
$\|\overrightarrow{BC}\| = \sqrt{(-2)^2 + 1^2} = \sqrt{5}$
$\|\overrightarrow{AC}\| = \sqrt{1^2 + (-1)^2} = \sqrt{2}$
Thus the length of the shortest side of S is $\sqrt{2}$.

(b) $\cos \angle BAC = \dfrac{\overrightarrow{AB} \cdot \overrightarrow{AC}}{\|\overrightarrow{AB}\| \cdot \|\overrightarrow{AC}\|} = \dfrac{2 \cdot 0 + 0 \cdot 1 + (-1) \cdot (-1)}{\sqrt{5} \cdot \sqrt{2}} \approx 0.32$

49. (a) The speed of the current is $\|\vec{c}\| = \sqrt{5} = 2.24$ m/sec.

(b) The speed of the current in the direction of the canoe's motion is the component of $\vec{c}$ in the direction of $\vec{v}$. This is given by:

$$\text{Speed of current in direction of canoe's motion} = \frac{\vec{c} \cdot \vec{v}}{\|\vec{v}\|} = \frac{(1)(5) + (2)(3)}{\sqrt{5^2 + 3^2}}$$
$$= \frac{11}{\sqrt{34}}$$
$$= 1.89 \text{ m/sec.}$$

Notice that the speed of the current in the direction of the canoe is less than the speed of the current in the direction in which the current is moving.

53. Let $\vec{u} = 3\vec{i} + 4\vec{j}$ and $\vec{v} = 5\vec{i} - 12\vec{j}$. We seek a vector $\vec{w} = x\vec{i} + y\vec{j}$ such that the cosine of the angle between $\vec{u}$ and $\vec{w}$ equals the cosine of the angle between $\vec{v}$ and $\vec{w}$. Thus

$$\frac{\vec{u} \cdot \vec{w}}{\|\vec{u}\| \|\vec{w}\|} = \frac{\vec{v} \cdot \vec{w}}{\|\vec{v}\| \|\vec{w}\|}$$

or

$$\frac{3x + 4y}{5\sqrt{x^2 + y^2}} = \frac{5x - 12y}{13\sqrt{x^2 + y^2}}.$$

Simplifying, we have $x = -8y$. The vector we want is of the form $\vec{w} = -8y\vec{i} + y\vec{j}$, but should we take $y > 0$ or $y < 0$? The smaller of the two angles formed by $\vec{u}$ and $\vec{v}$ is between $0°$ and $180°$, and so $\vec{w}$ must make an acute angle with $\vec{u}$ and $\vec{v}$. If $y > 0$ then $\vec{u} \cdot \vec{w} = -20y < 0$ indicating an obtuse angle and if $y < 0$ then $\vec{u} \cdot \vec{w} = -20y > 0$ indicating an acute angle. We have $\vec{w} = -8y\vec{i} + y\vec{j}$ with $y < 0$. Thus $\vec{w}$ can be any positive multiple of the vector $8\vec{i} - \vec{j}$.

57. Since $\vec{u} \cdot \vec{w} = \vec{v} \cdot \vec{w}$, $(\vec{u} - \vec{v}) \cdot \vec{w} = 0$. This equality holds for any $\vec{w}$, so we can take $\vec{w} = \vec{u} - \vec{v}$. This gives

$$\|\vec{u} - \vec{v}\|^2 = (\vec{u} - \vec{v}) \cdot (\vec{u} - \vec{v}) = 0,$$

that is,

$$\|\vec{u} - \vec{v}\| = 0.$$

This implies $\vec{u} - \vec{v} = 0$, that is, $\vec{u} = \vec{v}$.

61. Let $\vec{u}$ and $\vec{v}$ be the displacement vectors from C to the other two vertices. Then

$$\begin{aligned}
c^2 &= \|\vec{u} - \vec{v}\|^2 \\
&= (\vec{u} - \vec{v}) \cdot (\vec{u} - \vec{v}) \\
&= \vec{u} \cdot \vec{u} - \vec{v} \cdot \vec{u} - \vec{u} \cdot \vec{v} + \vec{v} \cdot \vec{v} \\
&= \|\vec{u}\|^2 - 2\|u\|\|v\| \cos C + \|\vec{v}\|^2 \\
&= a^2 - 2ab \cos C + b^2
\end{aligned}$$

Solutions for Section 13.4

Exercises

1. $\vec{v} \times \vec{w} = \vec{k} \times \vec{j} = -\vec{i}$ (remember $\vec{i}, \vec{j}, \vec{k}$ are unit vectors along the axes, and you must use the right hand rule.)

5. $\vec{v} = 2\vec{i} - 3\vec{j} + \vec{k}$, and $\vec{w} = \vec{i} + 2\vec{j} - \vec{k}$

$$\vec{v} \times \vec{w} = \begin{vmatrix} \vec{i} & \vec{j} & \vec{k} \\ 2 & -3 & 1 \\ 1 & 2 & -1 \end{vmatrix} = \vec{i} + 3\vec{j} + 7\vec{k}$$

9.

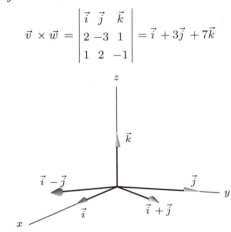

Figure 13.13

By definition, $(\vec{i} + \vec{j}) \times (\vec{i} - \vec{j})$ is in the direction of $-\vec{k}$. The magnitude is

$$\|\vec{i} + \vec{j}\| \cdot \|\vec{i} - \vec{j}\| \sin \frac{\pi}{2} = \sqrt{2} \cdot \sqrt{2} \cdot 1 = 2.$$

So $(\vec{i} + \vec{j}) \times (\vec{i} - \vec{j}) = -2\vec{k}$. See Figure 13.13.

13. The displacement vector from $(3, 4, 2)$ to $(-2, 1, 0)$ is:

$$\vec{a} = -5\vec{i} - 3\vec{j} - 2\vec{k}.$$

The displacement vector from $(3, 4, 2)$ to $(0, 2, 1)$ is:

$$\vec{b} = -3\vec{i} - 2\vec{j} - \vec{k}.$$

Therefore the vector normal to the plane is:

$$\vec{n} = \vec{a} \times \vec{b} = -\vec{i} + \vec{j} + \vec{k}.$$

Using the first point, the equation of the plane can be written as:

$$-(x - 3) + (y - 4) + (z - 2) = 0.$$

The equation of the plane is thus:

$$-x + y + z = 3.$$

Problems

17. We use the same normal $\vec{n} = 4\vec{i} + 26\vec{j} + 14\vec{k}$ and the point $(0, 0, 0)$ to get $4(x - 0) + 26(y - 0) + 14(z - 0) = 0$, or $4x + 26y + 14z = 0$.

21. The normal vectors to the planes are $\vec{n_1} = 2\vec{i} - 3\vec{j} + 5\vec{k}$ and $\vec{n_2} = 4\vec{i} + \vec{j} - 3\vec{k}$. The line of intersection is perpendicular to both normal vectors (picture the pages in a partially open book). Hence the vector we need is $\vec{n_1} \times \vec{n_2} = 4\vec{i} + 26\vec{j} + 14\vec{k}$.

25. (a) If we let $\overrightarrow{PQ}$ in Figure 13.14 be the vector from point P to point Q and $\overrightarrow{PR}$ be the vector from P to R, then

$$\overrightarrow{PQ} = -\vec{i} + 2\vec{k}$$

$$\overrightarrow{PR} = 2\vec{i} - \vec{k},$$

then the area of the parallelogram determined by $\overrightarrow{PQ}$ and $\overrightarrow{PR}$ is:

$$\begin{array}{c} \text{Area of} \\ \text{parallelogram} \end{array} = \|\overrightarrow{PQ} \times \overrightarrow{PR}\| = \left\| \begin{vmatrix} \vec{i} & \vec{j} & \vec{k} \\ -1 & 0 & 2 \\ 2 & 0 & -1 \end{vmatrix} \right\| = \|3\vec{j}\| = 3.$$

Thus, the area of the triangle PQR is

$$\left(\begin{array}{c} \text{Area of} \\ \text{triangle} \end{array} \right) = \frac{1}{2} \left(\begin{array}{c} \text{Area of} \\ \text{parallelogram} \end{array} \right) = \frac{3}{2} = 1.5.$$

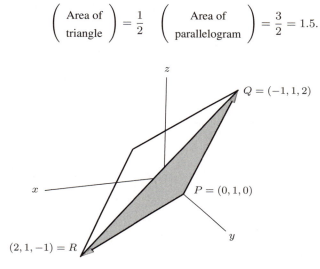

Figure 13.14

(b) Since $\vec{n} = \overrightarrow{PQ} \times \overrightarrow{PR}$ is perpendicular to the plane PQR, and from above, we have $\vec{n} = 3\vec{j}$, the equation of the plane has the form $3y = C$. At the point $(0, 1, 0)$ we get $3 = C$, therefore $3y = 3$, i.e., $y = 1$.

29. Since

$$\|\vec{v} \times \vec{w}\| = \|\vec{v}\| \cdot \|\vec{w}\| \sin\theta,$$

and

$$\vec{v} \cdot \vec{w} = \|\vec{v}\| \cdot \|\vec{w}\| \cos\theta,$$

so

$$\frac{\|\vec{v} \times \vec{w}\|}{\vec{v} \cdot \vec{w}} = \frac{\|\vec{v}\| \cdot \|\vec{w}\| \sin\theta}{\|\vec{v}\| \cdot \|\vec{w}\| \cos\theta} = \tan\theta,$$

so

$$\tan\theta = \frac{\|2\vec{i} - 3\vec{j} + 5\vec{k}\|}{3} = \frac{\sqrt{38}}{3} = 2.055.$$

33. The quantities $\left| \vec{a} \cdot (\vec{b} \times \vec{c}) \right|$ and $\left| (\vec{a} \times \vec{b}) \cdot \vec{c} \right|$ both represent the volume of the same parallelepiped, namely that defined by the three vectors $\vec{a}$, $\vec{b}$, and $\vec{c}$, and therefore must be equal. Thus, the two triple products $\vec{a} \cdot (\vec{b} \times \vec{c})$ and $(\vec{a} \times \vec{b}) \cdot \vec{c}$ must be equal except perhaps for their sign. In fact, both are positive if $\vec{a}$, $\vec{b}$, $\vec{c}$ are right-handed and negative if $\vec{a}$, $\vec{b}$, $\vec{c}$ are left-handed. This can be shown by drawing a picture:

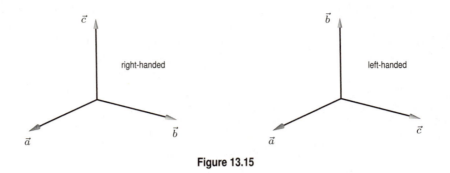

Figure 13.15

37. Write $\vec{v}$ and $\vec{w}$ in components and expand using the distributive property of the cross product.

$$\vec{v} \times \vec{w} = (v_1 \vec{i} + v_2 \vec{j} + v_3 \vec{k}) \times (w_1 \vec{i} + w_2 \vec{j} + w_3 \vec{k})$$
$$= v_1 w_1 \vec{i} \times \vec{i} + v_1 w_2 \vec{i} \times \vec{j} + v_1 w_3 \vec{i} \times \vec{k}$$
$$+ v_2 w_1 \vec{j} \times \vec{i} + v_2 w_2 \vec{j} \times \vec{j} + v_2 w_3 \vec{j} \times \vec{k}$$
$$+ v_3 w_1 \vec{k} \times \vec{i} + v_3 w_2 \vec{k} \times \vec{j} + v_3 w_3 \vec{k} \times \vec{k}$$

Now we use the fact that $\vec{i} \times \vec{i} = \vec{0}, \vec{i} \times \vec{j} = \vec{k}, \vec{i} \times \vec{k} = -\vec{j}, \vec{j} \times \vec{i} = -\vec{k}, \vec{j} \times \vec{j} = \vec{0}, \vec{j} \times \vec{k} = \vec{i}, \vec{k} \times \vec{i} = \vec{j}, \vec{k} \times \vec{j} = -\vec{i}, \vec{k} \times \vec{k} = \vec{0}$. Thus we have

$$\vec{v} \times \vec{w} = \vec{0} + v_1 w_2 \vec{k} + v_1 w_3 (-\vec{j}) + v_2 w_1 (-\vec{k}) + \vec{0} + v_2 w_3 \vec{i} + v_3 w_1 \vec{j} + v_3 w_2 (-\vec{i}) + \vec{0}$$
$$= (v_2 w_3 - v_3 w_2) \vec{i} + (v_3 w_1 - v_1 w_3) \vec{j} + (v_1 w_2 - v_2 w_1) \vec{k}.$$

41. The area vector for face $OAB = \frac{1}{2} \vec{b} \times \vec{a}$.
The area vector for face $OBC = \frac{1}{2} \vec{a} \times \vec{c}$.
The area vector for face $OAC = \frac{1}{2} \vec{b} \times \vec{c}$.
The area vector for face $ABC = \frac{1}{2} (\vec{b} - \vec{a}) \times (\vec{c} - \vec{a})$.

$$\frac{1}{2} \vec{b} \times \vec{a} + \frac{1}{2} \vec{c} \times \vec{b} + \frac{1}{2} \vec{a} \times \vec{c} + \frac{1}{2} (\vec{b} - \vec{a}) \times (\vec{c} - \vec{a}) =$$
$$\frac{1}{2} \vec{b} \times \vec{a} + \frac{1}{2} \vec{c} \times \vec{b} + \frac{1}{2} \vec{a} \times \vec{c} + \frac{1}{2} (\vec{b} \times \vec{c} - \vec{b} \times \vec{a} - \vec{a} \times \vec{c} - \vec{a} \times \vec{a}) = 0.$$

45. (a) Since

$$\vec{u} \times \vec{v} = (u_2 v_3 - u_3 v_2) \vec{i} + (u_3 v_1 - u_1 v_3) \vec{j} + (u_1 v_2 - u_2 v_1) \vec{k},$$

we have

$$\text{Area of } S = \| \vec{u} \times \vec{v} \| = \left((u_2 v_3 - u_3 v_2)^2 + (u_3 v_1 - u_1 v_3)^2 + (u_1 v_2 - u_2 v_1)^2 \right)^{1/2}.$$

(b) The two edges of R are given by the projections of $\vec{u}$ and $\vec{v}$ onto the xy-plane. These are the vectors $\vec{U}$ and $\vec{V}$, obtained by omitting the $\vec{k}$-components of $\vec{u}$ and $\vec{v}$: we have $\vec{U} = u_1 \vec{i} + u_2 \vec{j}$ and $\vec{V} = v_1 \vec{i} + v_2 \vec{j}$, Thus

$$\text{Area of } R = \| \vec{U} \times \vec{V} \| = \| (u_1 v_2 - u_2 v_1) \vec{k} \| = | u_1 v_2 - u_2 v_1 |.$$

(c) The vector $m \vec{i} + n \vec{j} - \vec{k}$ is normal to the plane $z = mx + ny + c$. Since the vectors $\vec{u}$ and $\vec{v}$ are in the plane (they're the sides of $\vec{S}$), the vector $\vec{u} \times \vec{v}$ is also normal to the plane. Thus, these two vectors are scalar multiples of one another. Suppose

$$\vec{u} \times \vec{v} = \lambda (m \vec{i} + n \vec{j} - \vec{k})$$

Since the $\vec{k}$ component of $\vec{u} \times \vec{v}$ is $(u_1v_2 - u_2v_1)\vec{k}$, comparing the $\vec{k}$-components tells us that

$$\lambda = -(u_1v_2 - u_2v_1).$$

Thus,

$$-(u_1v_2 - u_2v_1)(m\vec{i} + n\vec{j} - \vec{k}) = \vec{u} \times \vec{v} = (u_2v_3 - u_3v_2)\vec{i} + (u_3v_1 - u_1v_3)\vec{j} + (u_1v_2 - u_2v_1)\vec{k},$$

so

$$m = \frac{u_2v_3 - u_3v_2}{u_2v_1 - u_1v_2}$$
$$n = \frac{u_3v_1 - u_1v_3}{u_2v_1 - u_1v_2}.$$

(d) We have

$$(1 + m^2 + n^2) \cdot (\text{Area of } R)^2 = \left(1 + \left(\frac{u_2v_3 - u_3v_2}{u_2v_1 - u_1v_2}\right)^2 + \left(\frac{u_3v_1 - u_1v_3}{u_2v_1 - u_1v_2}\right)^2\right)(u_1v_2 - u_2v_1)^2$$
$$= (u_2v_3 - u_3v_2)^2 + (u_3v_1 - u_1v_3)^2 + (u_1v_2 - u_2v_1)^2$$
$$= (\text{Area of } S)^2.$$

Solutions for Chapter 13 Review

Exercises

1. Scalar. $\vec{u} \cdot \vec{v} = (2\vec{i} - 3\vec{j} - 4\vec{k}) \cdot (\vec{k} - \vec{j}) = 2 \cdot 0 - 3(-1) - 4 \cdot 1 = -1$.

5. $5\vec{c} = 5\vec{i} + 30\vec{j}$

9. $3\vec{v} - \vec{w} - \vec{v} = 2\vec{v} - \vec{w} = 2(2\vec{i} + 3\vec{j} - \vec{k}) - (\vec{i} - \vec{j} + 2\vec{k}) = 3\vec{i} + 7\vec{j} - 4\vec{k}$.

13. For any vector $\vec{v}$, we have $\vec{v} \times \vec{v} = \vec{0}$.

17. The cross product of two parallel vectors is $\vec{0}$, so the cross product of any vector with itself is $\vec{0}$.

21. (a) We have $\vec{v} \cdot \vec{w} = 3 \cdot 4 + 2 \cdot (-3) + (-2) \cdot 1 = 4$.

(b) We have $\vec{v} \times \vec{w} = -4\vec{i} - 11\vec{j} - 17\vec{k}$.

(c) A vector of length 5 parallel to $\vec{v}$ is

$$\frac{5}{\|\vec{v}\|}\vec{v} = \frac{5}{\sqrt{17}}(3\vec{i} + 2\vec{j} - 2\vec{k}) = 3.64\vec{i} + 2.43\vec{j} - 2.43\vec{k}.$$

(d) The angle between vectors $\vec{v}$ and $\vec{w}$ is found using

$$\cos\theta = \frac{\vec{v} \cdot \vec{w}}{\|\vec{v}\|\|\vec{w}\|} = \frac{4}{\sqrt{17}\sqrt{26}} = 0.190,$$

so $\theta = 79.0°$.

(e) The component of vector $\vec{v}$ in the direction of vector $\vec{w}$ is

$$\frac{\vec{v} \cdot \vec{w}}{\|\vec{w}\|} = \frac{4}{\sqrt{26}} = 0.784.$$

(f) The answer is any vector $\vec{a}$ such that $\vec{a} \cdot \vec{v} = 0$. One possible answer is $2\vec{i} - 2\vec{j} + \vec{k}$.

(g) A vector perpendicular to both is the cross product:

$$\vec{v} \times \vec{w} = -4\vec{i} - 11\vec{j} - 17\vec{k}.$$

25. $\vec{n} = 4\vec{i} + 6\vec{k}$ (the coefficients of x, y, z are the same as the coefficients of $\vec{i}, \vec{j}$, and $\vec{k}$.)

29. Since $\vec{F} = 2\vec{d}$, the two vectors are parallel in the same direction, so

$$\vec{F}_{\text{parallel}} = \vec{F} \text{ and } \vec{F}_{\text{perp}} = \vec{0}.$$

The work done is

$$W = \vec{F} \cdot \vec{d} = 2 + 8 = 10.$$

Notice that this is the same as the magnitude of the force, $\|\vec{F}\| = \sqrt{20}$, times the distance traveled, $\|\vec{d}\| = \sqrt{5}$, since the force is the same direction as the displacement.

33. The unit vector in the direction of $\vec{d}$ is $\vec{u} = (1/\sqrt{2})(\vec{i} + \vec{j})$. Thus

$$\vec{F}_{\text{parallel}} = \left(\vec{F} \cdot \vec{u}\right)\vec{u} = \frac{2}{\sqrt{2}}\vec{u} = \vec{i} + \vec{j},$$

$$\vec{F}_{\text{perp}} = \vec{F} - \vec{F}_{\text{parallel}} = \vec{i} - \vec{j}.$$

Notice that $\vec{F}_{\text{perp}} \cdot \vec{u} = 0$, as we expect. The work done is

$$W = \vec{F} \cdot \vec{d} = 2 - 0 = 2.$$

Problems

37. (a) The velocity vector for the boat is $\vec{b} = 25\vec{i}$ and the velocity vector for the current is

$$\vec{c} = -10\cos(45°)\vec{i} - 10\sin(45°)\vec{j} = -7.07\vec{i} - 7.07\vec{j}.$$

The actual velocity of the boat is

$$\vec{b} + \vec{c} = 17.93\vec{i} - 7.07\vec{j}.$$

(b) $\|\vec{b} + \vec{c}\| = 19.27$ km/hr.

(c) We see in Figure 13.16 that $\tan\theta = \dfrac{7.07}{17.93}$, so $\theta = 21.52°$ south of east.

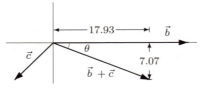

Figure 13.16

41. Vectors $\vec{v}_1$, $\vec{v}_4$, and $\vec{v}_8$ are all parallel to each other. Vectors $\vec{v}_3$, $\vec{v}_5$, and $\vec{v}_7$ are all parallel to each other, and are all perpendicular to the vectors in the previous sentence. Vectors $\vec{v}_2$ and $\vec{v}_9$ are perpendicular.

45. Since the plane is normal to the vector $5\vec{i} + \vec{j} - 2\vec{k}$ and passes through the point $(0, 1, -1)$, an equation for the plane is

$$5x + y - 2z = 5 \cdot 0 + 1 \cdot 1 + (-2) \cdot (-1) = 3$$
$$5x + y - 2z = 3.$$

49. (a) Since

$$\vec{PQ} = (3\vec{i} + 5\vec{j} + 7\vec{k}) - (\vec{i} + 2\vec{j} + 3\vec{k}) = 2\vec{i} + 3\vec{j} + 4\vec{k},$$

and

$$\vec{PR} = (2\vec{i} + 5\vec{j} + 3\vec{k}) - (\vec{i} + 2\vec{j} + 3\vec{k}) = \vec{i} + 3\vec{j},$$

$$\vec{PQ} \times \vec{PR} = \begin{vmatrix} \vec{i} & \vec{j} & \vec{k} \\ 2 & 3 & 4 \\ 1 & 3 & 0 \end{vmatrix} = -12\vec{i} + 4\vec{j} + 3\vec{k},$$

which is a vector perpendicular to the plane containing P, Q and R. Since

$$\|\vec{PQ} \times \vec{PR}\| = \sqrt{(-12)^2 + 4^2 + 3^2} = 13,$$

the unit vectors which are perpendicular to a plane containing P, Q, and R are

$$-\frac{12}{13}\vec{i} + \frac{4}{13}\vec{j} + \frac{3}{13}\vec{k},$$

or the unit vector pointing to the opposite direction,

$$\frac{12}{13}\vec{i} - \frac{4}{13}\vec{j} - \frac{3}{13}\vec{k}.$$

(b) The angle between PQ and PR is θ for which

$$\cos\theta = \frac{\overrightarrow{PQ}\cdot\overrightarrow{PR}}{\|\overrightarrow{PQ}\|\cdot\|\overrightarrow{PR}\|} = \frac{2\cdot 1 + 3\cdot 3 + 4\cdot 0}{\sqrt{2^2+3^2+4^2}\cdot\sqrt{1^2+3^2+0^2}} = \frac{11}{\sqrt{290}},$$

so

$$\theta = \cos^{-1}\left(\frac{11}{\sqrt{290}}\right) \approx 49.76°.$$

(c) The area of triangle $PQR = \frac{1}{2}\|\overrightarrow{PQ}\times\overrightarrow{PR}\| = \frac{13}{2}$.

(d) Let d be the distance from R to the line through P and Q (see Figure 13.17), then

$$\frac{1}{2}d\cdot\|\overrightarrow{PQ}\| = \text{the area of } \triangle PQR = \frac{13}{2}.$$

Therefore,

$$d = \frac{13}{\|\overrightarrow{PQ}\|} = \frac{13}{\sqrt{2^2+3^2+4^2}} = \frac{13}{\sqrt{29}}.$$

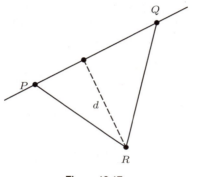

Figure 13.17

53. (a) On the x-axis, $y = z = 0$, so $5x = 21$, giving $x = \frac{21}{5}$. So the only such point is $\left(\frac{21}{5}, 0, 0\right)$.

(b) Other points are $(0, -21, 0)$, and $(0, 0, 3)$. There are many other possible answers.

(c) $\vec{n} = 5\vec{i} - \vec{j} + 7\vec{k}$. It is the normal vector.

(d) The vector between two points in the plane is parallel to the plane. Using the points from part (b), the vector $3\vec{k} - (-21\vec{j}) = 21\vec{j} + 3\vec{k}$ is parallel to the plane.

57. (a) Suppose $\vec{v} = \overrightarrow{OP}$ as in Figure 13.18. The $\vec{i}$ component of $\overrightarrow{OP}$ is the projection of $\overrightarrow{OP}$ on the x-axis:

$$\overrightarrow{OT} = v\cos\alpha\vec{i}.$$

Similarly, the $\vec{j}$ and $\vec{k}$ components of $\overrightarrow{OP}$ are the projections of $\overrightarrow{OP}$ on the y-axis and the z-axis respectively. So:

$$\overrightarrow{OS} = v\cos\beta\vec{j}$$
$$\overrightarrow{OQ} = v\cos\gamma\vec{k}$$

Since $\vec{v} = \overrightarrow{OT} + \overrightarrow{OS} + \overrightarrow{OQ}$, we have

$$\vec{v} = v\cos\alpha\vec{i} + v\cos\beta\vec{j} + v\cos\gamma\vec{k}.$$

(b) Since

$$\begin{aligned}
v^2 = \vec{v}\cdot\vec{v} &= (v\cos\alpha\vec{i} + v\cos\beta\vec{j} + v\cos\gamma\vec{k})\cdot \\
&\quad (v\cos\alpha\vec{i} + v\cos\beta\vec{j} + v\cos\gamma\vec{k}) \\
&= v^2(\cos^2\alpha + \cos^2\beta + \cos^2\gamma)
\end{aligned}$$

so

$$\cos^2\alpha + \cos^2\beta + \cos^2\gamma = 1.$$

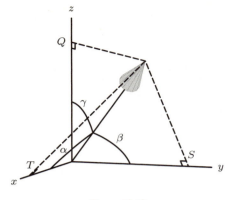

Figure 13.18

CAS Challenge Problems

61. (a) From the geometric definition of the dot product, we have

$$\cos \theta = \frac{|\vec{a} \cdot \vec{b}|}{\|\vec{a}\|\|\vec{b}\|} = \frac{10}{\sqrt{14}\sqrt{9}}.$$

Using $\sin^2 \theta = 1 - \cos^2 \theta$, we get

$$x + 2y + 3z = 0$$
$$2x + y + 2z = 0$$
$$x^2 + y^2 + z^2 = \|\vec{a}\|^2\|\vec{b}\|^2(1 - \cos^2 \theta) = (14)(9)\left(1 - \frac{100}{(14)(9)}\right)$$

Solving these equations we get $x = -1$, $y = -4$, $z = 3$ or $x = 1$, $y = 4$, and $z = -3$. Thus $\vec{c} = -\vec{i} - 4\vec{j} + 3\vec{k}$ or $\vec{c} = \vec{i} + 4\vec{j} - 3\vec{k}$.

(b) $\vec{a} \times \vec{b} = \vec{i} + 4\vec{j} - 3\vec{k}$. This is the same as one of the answers in part (a). The conditions in part (a) ensured that $\vec{c}$ is perpendicular to $\vec{a}$ and $\vec{b}$ and that it has magnitude $\|\vec{a}\|\|\vec{b}\||\sin \theta|$. The cross product is the solution that, in addition, satisfies the right-hand rule.

CHECK YOUR UNDERSTANDING

1. False. There are exactly two unit vectors: one in the same direction as $\vec{v}$ and the other in the opposite direction. Explicitly, the unit vectors parallel to $\vec{v}$ are $\pm\dfrac{1}{\|\vec{v}\|}\vec{v}$.

5. False. If $\vec{v}$ and $\vec{w}$ are not parallel, the three vectors $\vec{v}$, $\vec{w}$ and $\vec{v} - \vec{w}$ can be thought of as three sides of a triangle. (If the tails of $\vec{v}$ and $\vec{w}$ are placed together, then $\vec{v} - \vec{w}$ is a vector from the head of $\vec{w}$ to the head of $\vec{v}$.) The length of one side of a triangle is less than the sum of the lengths of the other two sides. Alternatively, a counterexample is $\vec{v} = \vec{i}$ and $\vec{w} = \vec{j}$. Then $\|\vec{i} - \vec{j}\| = \sqrt{2}$ but $\|\vec{i}\| - \|\vec{j}\| = 0$.

9. False. To find the displacement vector *from* $(1, 1, 1)$ *to* $(1, 2, 3)$ we subtract $\vec{i} + \vec{j} + \vec{k}$ from $\vec{i} + 2\vec{j} + 3\vec{k}$ to get $(1 - 1)\vec{i} + (2 - 1)\vec{j} + (3 - 1)\vec{k} = \vec{j} + 2\vec{k}$.

13. True. The cosine of the angle between the vectors is negative when the angle is between $\pi/2$ and π.

17. False. If the vectors are nonzero and perpendicular, the dot product will be zero (e.g. $\vec{i} \cdot \vec{j} = 0$).

21. True. The cross product yields a vector.

25. False. If $\vec{u}$ and $\vec{w}$ are two different vectors both of which are parallel to $\vec{v}$, then $\vec{v} \times \vec{u} = \vec{v} \times \vec{w} = \vec{0}$, but $\vec{u} \neq \vec{w}$. A counterexample is $\vec{v} = \vec{i}$, $\vec{u} = 2\vec{i}$ and $\vec{w} = 3\vec{i}$.

29. True. Any vector $\vec{w}$ that is parallel to $\vec{v}$ will give $\vec{v} \times \vec{w} = \vec{0}$.

CHAPTER FOURTEEN

Solutions for Section 14.1

Exercises

1. If h is small, then

$$f_x(3, 2) \approx \frac{f(3 + h, 2) - f(3, 2)}{h}.$$

With $h = 0.01$, we find

$$f_x(3, 2) \approx \frac{f(3.01, 2) - f(3, 2)}{0.01} = \frac{\frac{3.01^2}{(2+1)} - \frac{3^2}{(2+1)}}{0.01} = 2.00333.$$

With $h = 0.0001$, we get

$$f_x(3, 2) \approx \frac{f(3.0001, 2) - f(3, 2)}{0.0001} = \frac{\frac{3.0001^2}{(2+1)} - \frac{3^2}{(2+1)}}{0.0001} = 2.0000333.$$

Since the difference quotient seems to be approaching 2 as h gets smaller, we conclude

$$f_x(3, 2) \approx 2.$$

To estimate $f_y(3, 2)$, we use

$$f_y(3, 2) \approx \frac{f(3, 2 + h) - f(3, 2)}{h}.$$

With $h = 0.01$, we get

$$f_y(3, 2) \approx \frac{f(3, 2.01) - f(3, 2)}{0.01} = \frac{\frac{3^2}{(2.01+1)} - \frac{3^2}{(2+1)}}{0.01} = -0.99668.$$

With $h = 0.0001$, we get

$$f_y(3, 2) \approx \frac{f(3, 2.0001) - f(3, 2)}{0.0001} = \frac{\frac{3^2}{(2.0001+1)} - \frac{3^2}{(2+1)}}{0.0001} = -0.9999667.$$

Thus, it seems that the difference quotient is approaching -1, so we estimate

$$f_y(3, 2) \approx -1.$$

5. $\partial P / \partial t$: The unit is dollars per month. This is the rate at which payments change as the number of months it takes to pay off the loan changes. The sign is negative because payments decrease as the pay-off time increases.

$\partial P / \partial r$: The unit is dollars per percentage point. This is the rate at which payments change as the interest rate changes. The sign is positive because payments increase as the interest rate increases.

9. Moving right from P in the direction of increasing x increases f, so $f_x(P) > 0$.

Moving up from P in the direction of increasing y increases f, so $f_y(P) > 0$.

13. For $f_w(10, 25)$ we get

$$f_w(10, 25) \approx \frac{f(10 + h, 25) - f(10, 25)}{h}.$$

Choosing $h = 5$ and reading values from Table 12.2 on page 644 of the text, we get

$$f_w(10, 25) \approx \frac{f(15, 25) - f(10, 25)}{5} = \frac{13 - 15}{5} = -0.4°\text{F/mph}$$

This means that when the wind speed is 10 mph and the true temperature is $25°\text{F}$, as the wind speed increases from 10 mph by 1 mph we feel an approximately $0.4°\text{F}$ drop in temperature. This rate is negative because the temperature you feel drops as the wind speed increases.

Problems

17. The values of z increase as we move in the direction of increasing x-values, so f_x is positive. The values of z decrease as we move in the direction of increasing y-values, so f_y is negative. We see in the contour diagram that $f(2, 1) = 10$. We estimate the partial derivatives:

$$f_x(2, 1) \approx \frac{\Delta z}{\Delta x} = \frac{14 - 10}{4 - 2} = 2,$$

$$f_y(2, 1) \approx \frac{\Delta z}{\Delta y} = \frac{6 - 10}{2 - 1} = -4.$$

21. (a) (i) Near A, the value of z increases as x increases, so $f_x(A) > 0$.

(ii) Near A, the value of z decreases as y increases, so $f_y(A) < 0$.

(b) $f_x(P)$ changes from positive to negative as P moves from A to B along a straight line, because after P crosses the y-axis, z decreases as x increases near P.

$f_y(P)$ does not change sign as P moves from A to B along a straight line; it is negative along AB.

25. (a) Estimating $T(x, t)$ from the figure in the text at $x = 15$, $t = 20$ gives

$$\left.\frac{\partial T}{\partial x}\right|_{(15,20)} \approx \frac{T(23, 20) - T(15, 20)}{23 - 15} = \frac{20 - 23}{8} = -\frac{3}{8} \text{ °C per m,}$$

$$\left.\frac{\partial T}{\partial t}\right|_{(15,20)} \approx \frac{T(15, 25) - T(15, 20)}{25 - 20} = \frac{25 - 23}{5} = \frac{2}{5} \text{ °C per min.}$$

At 15 m from heater at time $t = 20$ min, the room temperature decreases by approximately $3/8$°C per meter and increases by approximately $2/5$°C per minute.

(b) We have the estimates,

$$\left.\frac{\partial T}{\partial x}\right|_{(5,12)} \approx \frac{T(7, 12) - T(5, 12)}{7 - 5} = \frac{25 - 27}{2} = -1 \text{ °C per m,}$$

$$\left.\frac{\partial T}{\partial t}\right|_{(5,12)} \approx \frac{T(5, 40) - T(5, 12)}{40 - 12} = \frac{30 - 27}{28} = \frac{3}{28} \text{ °C per min.}$$

At $x = 5$, $t = 12$ the temperature decreases by approximately 1°C per meter and increases by approximately $3/28$°C per minute.

29. (a) $\dfrac{\partial p}{\partial c} = f_c(c, s) = $ rate of change in blood pressure as cardiac output increases while systemic vascular resistance remains constant.

(b) Suppose that $p = kcs$. Note that c (cardiac output), a volume, s (SVR), a resistance, and p, a pressure, must all be positive. Thus k must be positive, and our level curves should be confined to the first quadrant. Several level curves are shown in Figure 14.1. Each level curve represents a different blood pressure level. Each point on a given curve is a combination of cardiac output and SVR that results in the blood pressure associated with that curve.

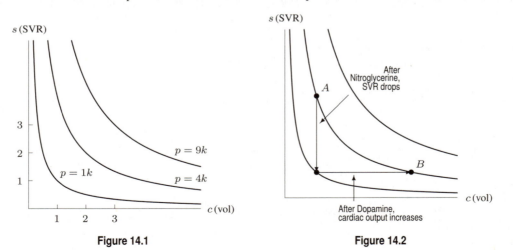

Figure 14.1   Figure 14.2

(c) Point B in Figure 14.2 shows that if the two doses are correct, the changes in pressure will cancel. The patient's cardiac output will have increased and his SVR will have decreased, but his blood pressure won't have changed.

(d) At point F in Figure 14.3, the patient's blood pressure is normalized, but his/her cardiac output has dropped and his SVR is up.

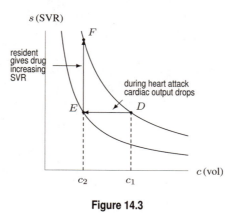

Figure 14.3

Note: c_1 and c_2 are the cardiac outputs before and after the heart attack, respectively.

Solutions for Section 14.2

Exercises

1. (a) Make a difference quotient using the two points $(3, 2)$ and $(3, 2.01)$ that have the same x-coordinate 3 but whose y-coordinates differ by 0.01. We have

$$f_y(3, 2) \approx \frac{f(3, 2.01) - f(3, 2)}{2.01 - 2} = \frac{28.0701 - 28}{0.01} = 7.01.$$

(b) Differentiating gives $f_y(x, y) = x + 2y$, so $f_y(3, 2) = 3 + 2 \cdot 2 = 7$.

5. $\dfrac{\partial z}{\partial x} = \dfrac{\partial}{\partial x}\left[(x^2 + x - y)^7\right] = 7(x^2 + x - y)^6(2x + 1) = (14x + 7)(x^2 + x - y)^6.$

$\dfrac{\partial z}{\partial y} = \dfrac{\partial}{\partial y}\left[(x^2 + x - y)^7\right] = -7(x^2 + x - y)^6.$

9. $\dfrac{\partial}{\partial T}\left(\dfrac{2\pi r}{T}\right) = -\dfrac{2\pi r}{T^2}$

13. $a_v = \dfrac{2v}{r}$

17. $\dfrac{\partial}{\partial r}\left(\dfrac{2\pi r}{v}\right) = \dfrac{2\pi}{v}$

21. $z_x = -\sin x, \quad z_x(2, 3) = -\sin 2 \approx -0.9$

25. $\dfrac{\partial}{\partial M}\left(\dfrac{2\pi r^{3/2}}{\sqrt{GM}}\right) = 2\pi r^{3/2}(-\dfrac{1}{2})(GM)^{-3/2}(G) = -\pi r^{3/2} \cdot \dfrac{G}{GM\sqrt{GM}} = -\dfrac{\pi r^{3/2}}{M\sqrt{GM}}$

29. $\dfrac{\partial V}{\partial r} = \dfrac{8}{3}\pi rh$ and $\dfrac{\partial V}{\partial h} = \dfrac{4}{3}\pi r^2.$

33. $z_x = 7x^6 + yx^{y-1}$, and $z_y = 2^y \ln 2 + x^y \ln x$

Problems

37. **(a)** The difference quotient for evaluating $f_w(2, 2)$ is

$$f_w(2, 2) \approx \frac{f(2 + 0.01, 2) - f(2, 2)}{h} = \frac{e^{(2.01) \ln 2} - e^{2 \ln 2}}{0.01} = \frac{e^{\ln(2^{2.01})} - e^{\ln(2^2)}}{0.01}$$

$$= \frac{2^{(2.01)} - 2^2}{0.01} \approx 2.78$$

The difference quotient for evaluating $f_z(2, 2)$ is

$$f_z(2, 2) \approx \frac{f(2, 2 + 0.01) - f(2, 2)}{h}$$

$$= \frac{e^{2 \ln(2.01)} - e^{2 \ln 2}}{0.01} = \frac{(2.01)^2 - 2^2}{0.01} = 4.01$$

(b) Using the derivative formulas we get

$$f_w = \frac{\partial f}{\partial w} = \ln z \cdot e^{w \ln z} = z^w \cdot \ln z$$

$$f_z = \frac{\partial f}{\partial z} = e^{w \ln z} \cdot \frac{w}{z} = w \cdot z^{w-1}$$

so

$$f_w(2, 2) = 2^2 \cdot \ln 2 \approx 2.773$$

$$f_z(2, 2) = 2 \cdot 2^{2-1} = 4.$$

41. **(a)** $\dfrac{\partial E}{\partial m} = c^2 \left(\dfrac{1}{\sqrt{1 - v^2/c^2}} - 1 \right)$. We expect this to be positive because energy increases with mass.

(b) $\dfrac{\partial E}{\partial v} = mc^2 \cdot \left(-\dfrac{1}{2} \right) (1 - v^2/c^2)^{-3/2} \left(-\dfrac{2v}{c^2} \right) = \dfrac{mv}{(1 - v^2/c^2)^{3/2}}$. We expect this to be positive because energy increases with velocity.

45. Since $f_x(x, y) = 4x^3 y^2 - 3y^4$, we could have

$$f(x, y) = x^4 y^2 - 3xy^4.$$

In that case,

$$f_y(x, y) = \frac{\partial}{\partial y} (x^4 y^2 - 3xy^4) = 2x^4 y - 12xy^3$$

as expected. More generally, we could have $f(x, y) = x^4 y^2 - 3xy^4 + C$, where C is any constant.

Solutions for Section 14.3

Exercises

1. The partial derivatives are

$$z_x = x \quad \text{and} \quad z_y = 4y,$$

so

$$z(2, 1) = 4, \quad z_x(2, 1) = 2 \quad \text{and} \quad z_y(2, 1) = 4.$$

The tangent plane to $z = \frac{1}{2}(x^2 + 4y^2)$ at $(x, y) = (2, 1)$ has equation

$$z = z(2, 1) + z_x(2, 1)(x - 2) + z_y(2, 1)(y - 1)$$

$$= 4 + 2(x - 2) + 4(y - 1)$$

$$= -4 + 2x + 4y.$$

5. Since

$$z_x = y \cos xy, \text{ we have } z_x \left(2, \frac{3\pi}{4}\right) = \frac{3\pi}{4} \cos \left(2 \cdot \frac{3\pi}{4}\right) = 0.$$

$$z_y = x \cos xy, \text{ we have } z_y \left(2, \frac{3\pi}{4}\right) = 2 \cos \left(2 \cdot \frac{3\pi}{4}\right) = 0.$$

Since

$$z \left(2, \frac{3\pi}{4}\right) = \sin \left(2 \cdot \frac{3\pi}{4}\right) = -1,$$

the tangent plane is

$$z = z \left(2, \frac{3\pi}{4}\right) + z_x \left(2, \frac{3\pi}{4}\right)(x - 2) + z_y \left(2, \frac{3\pi}{4}\right)\left(y - \frac{3\pi}{4}\right)$$

$$z = -1.$$

9. Since $g_u = 2u + v$ and $g_v = u$, we have

$$dg = (2u + v)\, du + u\, dv$$

13. We have $df = f_x\, dx + f_y\, dy$. Finding the partial derivatives, we have $f_x = e^{-y}$ so $f_x(1, 0) = e^{-0} = 1$, and $f_y = -xe^{-y}$ so $f_y(1, 0) = -1e^{-0} = -1$. Thus, $df = dx - dy$.

Problems

17. (a) The units are dollars/square foot.
  (b) The price of land 300 feet from the beach and of area near 1000 square feet is greater for larger plots by about $3 per square foot.
  (c) The units are dollars/foot.
  (d) The price of a 1000 square foot plot about 300 feet from the beach is less for plots farther from the beach by about $2 per extra foot from the beach.
  (e) Compared to the 998 ft^2 plot at 295 ft from the beach, the other plot costs about $7 \times 3 = \$21$ more for the extra 7 square feet but about $10 \times 2 = \$20$ less for the extra 10 feet you have to walk to the beach. The net difference is about a dollar, and the smaller plot nearer the beach is cheaper.

21. Since $f_x(x, y) = \frac{x}{\sqrt{x^2+y^3}}$ and $f_y(x, y) = \frac{3y^2}{2\sqrt{x^2+y^3}}$,
  $f_x(1, 2) = \frac{1}{\sqrt{1^2+2^3}} = \frac{1}{3}$ and $f_y(1, 2) = \frac{3 \cdot 2^2}{2\sqrt{1^2+2^3}} = 2$.
  Thus the differential at the point $(1, 2)$ is

$$df = df(1, 2) = f_x(1, 2)dx + f_y(1, 2)dy = \frac{1}{3}dx + 2dy.$$

Using the differential at the point $(1, 2)$, we can estimate $f(1.04, 1.98)$. Since

$$\triangle f \approx f_x(1, 2)\triangle x + f_y(1, 2)\triangle y$$

where $\triangle f = f(1.04, 1.98) - f(1, 2)$ and $\triangle x = 1.04 - 1$ and $\triangle y = 1.98 - 2$, we have

$$f(1.04, 1.98) \approx f(1, 2) + f_x(1, 2)(1.04 - 1) + f_y(1, 2)(1.98 - 2)$$

$$= \sqrt{1^2 + 2^3} + \frac{0.04}{3} - 2(0.02) \approx 2.973.$$

25. (a) The linear approximation gives

$$f(520, 24) \approx 24.20, \quad f(480, 24) \approx 23.18,$$

$$f(500, 22) \approx 25.52, \quad f(500, 26) \approx 21.86.$$

The approximations for $f(520, 24)$ and $f(500, 26)$ agree exactly with the values in the table; the other two do not. The reason for this is that the partial derivatives were estimated using difference quotients with these values.

(b) We could get a more balanced estimate by using a difference quotient that uses the values on both sides. Thus, we could estimate the partial derivatives as follows:

$$f_T(500, 24) \approx \frac{f(520, 24) - f(480, 24)}{40}$$

$$= \frac{(24.20 - 23.19)}{40} = 0.02525,$$

and

$$f_p(500, 24) \approx \frac{f(500, 26) - f(500, 22)}{4}$$

$$= \frac{(21.86 - 25.86)}{4} = -1.$$

This yields the linear approximation

$$V = f(T, p) \approx 23.69 + 0.02525(T - 500) - (p - 24) \text{ ft}^3.$$

This approximation yields values

$$f(520, 24) \approx 24.195, \quad f(480, 24) \approx 23.185,$$
$$f(500, 22) \approx 25.69, \quad f(500, 26) \approx 21.69.$$

Although none of these predictions are accurate, the error in the predictions that were wrong before has been reduced. This new linearization is a better all-round approximation for values near $(500, 24)$.

29. The error in η is approximated by $d\eta$, where

$$d\eta = \frac{\partial \eta}{\partial r} dr + \frac{\partial \eta}{\partial p} dp.$$

We need to find

$$\frac{\partial \eta}{\partial r} = \frac{\pi}{8} \frac{p 4 r^3}{v}$$

and

$$\frac{\partial \eta}{\partial p} = \frac{\pi}{8} \frac{r^4}{v}.$$

For $r = 0.005$ and $p = 10^5$ we get

$$\frac{\partial \eta}{\partial r}(0.005, 10^5) = 3.14159 \cdot 10^7, \quad \frac{\partial \eta}{\partial p}(0.005, 10^5) = 0.39270,$$

so that

$$d\eta = \frac{\partial \eta}{\partial r} dr + \frac{\partial \eta}{\partial p} dp.$$

is largest when we take all positive values to give

$$d\eta = 3.14159 \cdot 10^7 \cdot 0.00025 + 0.39270 \cdot 1000 = 8246.68.$$

This seems quite large but $\eta(0.005, 10^5) = 39269.9$ so the maximum error represents about 20% of any value computed by the given formula. Notice also the relative error in r is $\pm 5\%$, which means the relative error in r^4 is $\pm 20\%$.

Solutions for Section 14.4

Exercises

1. Since the partial derivatives are

$$\frac{\partial f}{\partial x} = \frac{15}{2} x^4 - 0 = \frac{15}{2} x^4$$

$$\frac{\partial f}{\partial y} = 0 - \frac{24}{7} y^5 = -\frac{24}{7} y^5$$

we have

$$\text{grad } f = \frac{\partial f}{\partial x} \vec{i} + \frac{\partial f}{\partial y} \vec{j} = \left(\frac{15}{2} x^4\right) \vec{i} - \left(\frac{24}{7} y^5\right) \vec{j}.$$

5. Since the partial derivatives are

$$z_x = e^y \quad \text{and} \quad z_y = xe^y + e^y + ye^y,$$

we have

$$\nabla z = e^y \vec{i} + e^y(1 + x + y)\vec{j}.$$

9. Since the partial derivatives are

$$f_r = \sin\theta \quad \text{and} \quad f_\theta = r\cos\theta,$$

we have

$$\nabla f = \sin\theta \vec{i} + r\cos\theta \vec{j}.$$

13. Since the partial derivatives are

$$\frac{\partial f}{\partial \alpha} = \frac{(2\alpha - 3\beta)(2 + 0) - (2 - 0)(2\alpha + 3\beta)}{(2\alpha - 3\beta)^2}$$

$$= \frac{4\alpha - 6\beta - (4\alpha + 6\beta)}{(2\alpha - 3\beta)^2}$$

$$= -\frac{12\beta}{(2\alpha - 3\beta)^2}$$

$$\frac{\partial f}{\partial \beta} = \frac{(2\alpha - 3\beta)(0 + 3) - (0 - 3)(2\alpha + 3\beta)}{(2\alpha - 3\beta)^2}$$

$$= \frac{(6\alpha - 9\beta) + (6\alpha + 9\beta)}{(2\alpha - 3\beta)^2}$$

$$= \frac{12\alpha}{(2\alpha - 3\beta)^2}$$

we have

$$\text{grad } f = \frac{\partial f}{\partial \alpha}\vec{i} + \frac{\partial f}{\partial \beta}\vec{j} = \left(-\frac{12\beta}{(2\alpha - 3\beta)^2}\right)\vec{i} + \left(\frac{12\alpha}{(2\alpha - 3\beta)^2}\right)\vec{j}.$$

17. Since the partial derivatives are

$$f_r = 2\pi(h + r) \quad \text{and} \quad f_h = 2\pi r,$$

we have

$$\nabla f(2, 3) = 10\pi \vec{i} + 4\pi \vec{j}.$$

21. Since the partial derivatives are

$$\frac{\partial f}{\partial x} = \frac{1}{2}(\tan x + y)^{-1/2}\left(\frac{1}{\cos^2 x} + 0\right) = \frac{1}{2\cos^2 x\sqrt{\tan x + y}},$$

and

$$\frac{\partial f}{\partial y} = \frac{1}{2}(\tan x + y)^{-1/2}(0 + 1) = \frac{1}{2\sqrt{\tan x + y}},$$

then

$$\text{grad } f = \frac{\partial f}{\partial x}\vec{i} + \frac{\partial f}{\partial y}\vec{j} = \left(\frac{1}{2\cos^2 x\sqrt{\tan x + y}}\right)\vec{i} + \left(\frac{1}{2\sqrt{\tan x + y}}\right)\vec{j}.$$

Hence we have

$$\text{grad } f\Big|_{(0,1)} = \left(\frac{1}{2(\cos(0))^2\sqrt{\tan(0) + 1}}\right)\vec{i} + \left(\frac{1}{2\sqrt{\tan(0) + 1}}\right)\vec{j}$$

$$= \left(\frac{1}{2(1)^2\sqrt{0 + 1}}\right)\vec{i} + \left(\frac{1}{2\sqrt{0 + 1}}\right)\vec{j}$$

$$= \frac{1}{2}\vec{i} + \frac{1}{2}\vec{j}$$

25. Since $f_x = 2\cos(2x - y)$ and $f_y = -\cos(2x - y)$, at $(1, 2)$ we have grad $f = 2\cos(0)\vec{i} - \cos(0)\vec{j} = 2\vec{i} - \vec{j}$. Thus

$$f_{\vec{u}}(1, 2) = \text{grad } f \cdot \left(\frac{3}{5}\vec{i} - \frac{4}{5}\vec{j}\right) = \frac{2 \cdot 3 - 1(-4)}{5} = \frac{10}{5} = 2.$$

29. Since $df = f_x dx + f_y dy$ and grad $f = f_x \vec{i} + f_y \vec{j}$ we have

$$\text{grad } f = (x + 1)ye^x \vec{i} + xe^x \vec{j}.$$

33. Since the values of z increase as we move in direction $-\vec{i} + \vec{j}$ from the point $(-1, 1)$, the directional derivative is positive.

37. The approximate direction of the gradient vector at point $(0, 2)$ is $\vec{j}$, since the gradient vector is perpendicular to the contour and points in the direction of increasing z-values. Answers may vary since answers are approximate and any positive multiple of the vector given is also correct.

41. The approximate direction of the gradient vector at point $(2, -2)$ is $\vec{i} - \vec{j}$, since the gradient vector is perpendicular to the contour and points in the direction of increasing z-values. Answers may vary since answers are approximate and any positive multiple of the vector given is also correct.

Problems

45. (a) The unit vector $\vec{u}$ in the same direction as $\vec{v}$ is

$$\vec{u} = \frac{1}{\|\vec{v}\|}\vec{v} = \frac{1}{\sqrt{(-1)^2 + 3^2}}\vec{v} = \frac{-1}{\sqrt{10}}\vec{i} + \frac{3}{\sqrt{10}}\vec{j} = -0.316228\vec{i} + 0.948683\vec{j}.$$

The vector $\vec{w}$ of length 0.1 in the direction of $\vec{v}$ is

$$\vec{w} = 0.1\vec{u} = -0.0316228\vec{i} + 0.0948683\vec{j}.$$

The displacement vector from P to Q is $\vec{w}$. Hence

$$Q = (4 - 0.0316228, 5 + 0.0948683) = (3.96838, 5.09487).$$

(b) Since the distance from P to Q is 0.1, the directional derivative of f at P in the direction of Q is approximately

$$f_{\vec{u}} \approx \frac{f(Q) - f(P)}{0.1} = \frac{3.01052 - 3}{0.1} = 0.1052.$$

(c) We have

$$\text{grad } f(x, y) = \frac{1}{2\sqrt{x + y}}\vec{i} + \frac{1}{2\sqrt{x + y}}\vec{j}$$

$$\text{grad } f(4, 5) = \frac{1}{6}\vec{i} + \frac{1}{6}\vec{j}.$$

The directional derivative at $P = (4, 5)$ in the direction of $\vec{u}$ is

$$f_{\vec{u}}(4, 5) = \text{grad } f(4, 5) \cdot \vec{u} = \frac{1}{6}\frac{-1}{\sqrt{10}} + \frac{1}{6}\frac{3}{\sqrt{10}} = \frac{1}{3\sqrt{10}} = 0.1054.$$

49. At points (x, y) where the gradients are defined and are not the zero vector, the level curves of f and g intersect at right angles if and only grad $f \cdot$ grad $g = 0$.

We have grad $f \cdot$ grad $g = (\vec{i} + \vec{j}) \cdot (\vec{i} - \vec{j}) = 0$. The level curves of f and g are straight lines that cross at right angles. See Figure 14.4.

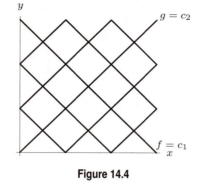

Figure 14.4

53. $f_{\vec{i}}(3, 1)$ means the rate of change of f in the x direction at $(3, 1)$. Thus,

$$f_{\vec{i}}(3, 1) \approx \frac{f(4, 1) - f(3, 1)}{1} = \frac{2 - 1}{1} = 1.$$

57. One way to do this is to estimate the gradient vector and then find grad $f(x, y) \cdot \vec{u}$. This is a useful approach since it is easier to estimate grad f than to estimate $f_{\vec{u}}$ directly. Since grad $f(x, y) = (f_x(x, y), f_y(x, y))$ we can simply estimate the x- and y-derivatives of f at $(3, 1)$ to find grad f at that point. In the x-direction we see that the function is increasing. This implies that the x-derivative is positive. To estimate its value we estimate the slope in the x-direction. Applying the same reasoning to find f_y, we get:

$$f_x(3, 1) \approx \frac{f(3 + 1, 1) - f(3, 1)}{1} = 2 - 1 = 1,$$
$$f_y(3, 1) \approx \frac{f(3, 1) - f(3, 1 - 0.6)}{0.6} \approx \frac{1 - 2}{0.6} \approx -1.67.$$

This gives us our estimated value for grad $f(3, 1) \approx \vec{i} - 1.67\vec{j}$. Now, if $\vec{u} = (-2\vec{i} + \vec{j})/\sqrt{5}$,

$$f_{\vec{u}}(3, 1) = \text{grad } f(3, 1) \cdot (-2\vec{i} + \vec{j})/\sqrt{5}$$
$$\approx (\vec{i} - 1.67\vec{j}) \cdot (-2\vec{i} + \vec{j})/\sqrt{5} \approx -1.64$$

61. First, we check that $3 = 2^2 - 1$. Then let $f(x, y) = y - x^2 + 1$ so that the given curve is the contour $f(x, y) = 0$. Since $f_x = -2x$ and $f_y = 1$, we have grad $f(2, 3) = -4\vec{i} + \vec{j}$. Since gradients are perpendicular to contours, a vector normal to the curve at $(2, 3)$ is $\vec{n} = -4\vec{i} + 1\vec{j}$. Using the normal vector to a line the same way we use the normal vector to a plane, we get that an equation of the tangent line is $-4(x - 2) + (y - 3) = 0$. Notice, if we had instead found the slope of the tangent line using $dy/dx = 2x$, we get $(y - 3) = 4(x - 2)$, which agrees with the equation we got using the gradient.

65. (a) Negative. ∇f is perpendicular to the level curve at the point P, so its x-component which is $\nabla f \cdot \vec{i}$ is negative.
(b) Positive. The y-component of ∇f is in the same direction as $\vec{j}$ at P and hence the dot product will be positive.
(c) Positive. The partial derivative with respect to x at Q is positive because the value of f is increasing in the positive x direction at Q. (Note that Q lies between the level curves with values 3 and 4 and that the one with value 4 is further in the positive x direction from Q.)
(d) Negative. Again, Q lies between the level curves with values 3 and 4 and the one with value 3 is further from Q in the positive y direction, so the partial derivative with respect to y at Q is negative.

69. We see that

$$\text{grad } f = (3y)\vec{i} + (3x + 2y)\vec{j},$$

so at the point $(2, 3)$, we have

$$\text{grad } f = 9\vec{i} + 12\vec{j}.$$

(a) The directional derivative is $\nabla f \cdot \dfrac{\vec{v}}{\|\vec{v}\|} = \dfrac{(9)(3) + (12)(-1)}{\sqrt{10}} = \dfrac{15}{\sqrt{10}} \approx 4.74.$
(b) The direction of maximum rate of change is $\nabla f(2, 3) = 9\vec{i} + 12\vec{j}$.
(c) The maximum rate of change is $\|\nabla f\| = \sqrt{225} = 15$.

73. Directional derivative $= \nabla f \cdot \vec{u}$, where $\vec{u} =$ unit vector. If we move from $(4, 5)$ to $(5, 6)$, we move in the direction $\vec{i} + \vec{j}$ so $\vec{u} = \frac{1}{\sqrt{2}}\vec{i} + \frac{1}{\sqrt{2}}\vec{j}$. So,

$$\nabla f \cdot \vec{u} = f_x \left(\frac{1}{\sqrt{2}} \right) + f_y \left(\frac{1}{\sqrt{2}} \right) = 2.$$

Similarly, if we move from $(4, 5)$ to $(6, 6)$, the direction is $2\vec{i} + \vec{j}$ so $\vec{u} = \frac{2}{\sqrt{5}}\vec{i} + \frac{1}{\sqrt{5}}\vec{j}$. So

$$\nabla f \cdot \vec{u} = f_x \left(\frac{2}{\sqrt{5}} \right) + f_y \left(\frac{1}{\sqrt{5}} \right) = 3.$$

Solving the system of equations for f_x and f_y

$$f_x + f_y = 2\sqrt{2}$$
$$2f_x + f_y = 3\sqrt{5}$$

gives

$$f_x = 3\sqrt{5} - 2\sqrt{2}$$
$$f_y = 4\sqrt{2} - 3\sqrt{5}.$$

Thus at $(4, 5)$,

$$\nabla f = (3\sqrt{5} - 2\sqrt{2})\vec{i} + (4\sqrt{2} - 3\sqrt{5})\vec{j}.$$

77. (a) If $\vec{j}$ points north and $\vec{i}$ points east, then the direction the car is driving is $\vec{j} - \vec{i}$. A unit vector in this direction is

$$\vec{u} = \frac{1}{\sqrt{2}}(\vec{j} - \vec{i}).$$

The gradient of the height function is

$$\operatorname{grad} h = \frac{\partial h}{\partial E}\vec{i} + \frac{\partial h}{\partial N}\vec{j} = 50\vec{i} + 100\vec{j}.$$

So the directional derivative is

$$h_{\vec{u}} = \operatorname{grad} g \cdot \vec{h} = (50\vec{i} + 100\vec{j}) \cdot \frac{1}{\sqrt{2}}(\vec{j} - \vec{i}) = \frac{100 - 50}{\sqrt{2}} = 35.355 \text{ ft/mi.}$$

(b) The car is traveling at v mi/hr, so

$$\text{Rate of change of with respect to time} = v \frac{\text{miles}}{\text{hour}} 35.355 \frac{\text{ft}}{\text{mile}} = 35.355v \text{ ft/hour.}$$

81. (a) To estimate the change in f, we use the gradient vector to estimate the change in f in moving from P to Q. Because the contours are approximately parallel, moving from P to Q takes you to the same contour as moving from P to R. (See Figure 14.5.) If θ is the angle between $\vec{u}$ and $\operatorname{grad} f(a, b)$, then

$$\begin{array}{l} \text{Change in } f \\ \text{between } P \text{ and } Q \end{array} = \text{Change in } f$$

$$= \left(\begin{array}{c} \text{Rate of change} \\ \text{in direction } PR \end{array} \right) \left(\begin{array}{c} \text{Distance traveled} \\ \text{between } P \text{ and } R \end{array} \right)$$

$$\approx \| \operatorname{grad} f \| (h \cos \theta).$$

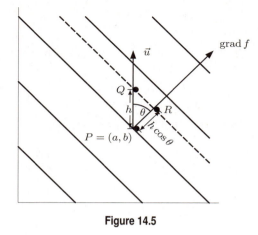

Figure 14.5

(b) Since $\vec{u}$ is a unit vector, we use the definition of $f_{\vec{u}}(a, b)$ to estimate

$$f_{\vec{u}}(a, b) \approx \frac{\text{Change in } f}{h} \approx \frac{\|\operatorname{grad} f(a, b)\| h \cos\theta}{h}$$
$$= \|\operatorname{grad} f(a, b)\| \cos\theta = \|\operatorname{grad} f\| \|\vec{u}\| \cos\theta = \operatorname{grad} f(a, b) \cdot \vec{u}.$$

This approximation gets better as we choose h smaller and smaller, and in the limit we get the formula:

$$f_{\vec{u}}(a, b) = \operatorname{grad} f(a, b) \cdot \vec{u}.$$

Solutions for Section 14.5

Exercises

1. Since $f_x = 2x$, $f_y = 0$ and $f_z = 0$, we have

$$\operatorname{grad} f = 2x\vec{i}.$$

5. Since $f(x, y, z) = \dfrac{1}{xyz}$, we have

$$f_x = -\frac{1}{x^2yz}, \quad f_y = -\frac{1}{xy^2z} \quad f_z = -\frac{1}{xyz^2},$$

we have

$$\operatorname{grad} f = -\frac{1}{xyz}\left(\frac{1}{x}\vec{i} + \frac{1}{y}\vec{j} + \frac{1}{z}\vec{k}\right).$$

9. We have $f_x = e^y \sin z$, $f_y = xe^y \sin z$, and $f_z = xe^y \cos z$. Thus

$$\operatorname{grad} f = e^y \sin z\vec{i} + xe^y \sin z\vec{j} + xe^y \cos z\vec{k}.$$

13. We have $f_x = 0$, $f_y = 2yz$ and $f_z = y^2$. Thus $\operatorname{grad} f = 2yz\vec{j} + y^2\vec{k}$ and $\operatorname{grad} f(1, 0, 1) = \vec{0}$.

17. We have

$$\operatorname{grad}(x\ln(yz))\bigg|_{(2,1,e)} = \ln(yz)\vec{i} + \frac{x}{y}\vec{j} + \frac{x}{z}\vec{k}\bigg|_{(2,1,e)} = \vec{i} + 2\vec{j} + \frac{2}{e}\vec{k}.$$

21. We have grad $f = y\vec{i} + x\vec{j} + 2z\vec{k}$, so grad $f(1,1,0) = \vec{i} + \vec{j}$. A unit vector in the direction we want is $u = (1/\sqrt{2})(-\vec{i} + \vec{k})$. Therefore, the directional derivative is

$$\text{grad } f(1,1,0) \cdot \vec{u} = \frac{1(-1) + 1 \cdot 0 + 0 \cdot 1}{\sqrt{2}} = \frac{-1}{\sqrt{2}}.$$

25. First, we check that $(-1)^2 - (1)^2 + 2^2 = 4$. Then let $f(x,y,z) = x^2 - y^2 + z^2$ so that the given surface is the level surface $f(x,y,z) = 4$. Since $f_x = 2x$, $f_y = -2y$, and $f_z = 2z$, we have grad $f(-1,1,2) = -2\vec{i} - 2\vec{j} + 4\vec{k}$. Since gradients are perpendicular to level surfaces, a vector normal to the surface at $(-1,1,2)$ is $\vec{n} = -2\vec{i} - 2\vec{j} + 4\vec{k}$. Thus an equation for the tangent plane is

$$-2(x + 1) - 2(y - 1) + 4(z - 2) = 0.$$

29. First, we check that $\cos(-1 + 1) = e^{-2+2}$. Then we let $f(x,y,z) = \cos(x + y) - e^{xz+2}$, so that the given surface is the level surface $f(x,y,z) = 0$. Since $f_x = -\sin(x + y) - ze^{xz+2}$, $f_y = -\sin(x + y)$, and $f_z = -xe^{xz+2}$, we have grad $f(-1,1,2) = -2\vec{i} + \vec{k}$. Since gradients are perpendicular to level surfaces, a vector normal to the surface at $(-1,1,2)$ is $\vec{n} = -2\vec{i} + \vec{k}$. Thus an equation for the tangent plane is

$$-2(x + 1) + (z - 2) = 0.$$

33. (a) We have $f_x = 2x - yz$, $f_y = 2y - xz$ and $f_z = -xy$ so

$$\text{grad } f = (2x - yz)\vec{i} + (2y - xz)\vec{j} - xy\vec{k}.$$

(b) At the point $(2, 3, 1)$ we have

$$\text{grad } f(2,3,1) = \vec{i} + 4\vec{j} + 6\vec{k}.$$

Thus an equation of the tangent plane to the level surface at the point $(2, 3, 1)$ is

$$(x - 2) + 4(y - 3) + 6(z - 1) = 0$$

or

$$x + 4y + 6z = 20.$$

37. Let $f(x,y,z) = x^2 + y^2$ so that the surface is the level surface $f(x,y,z) = 1$. Since

$$\text{grad } f = 2x\vec{i} + 2y\vec{j}$$

we have

$$\text{grad } f(1,0,1) = 2\vec{i}.$$

Thus an equation of the tangent plane at the point $(1, 0, 1)$ is

$$2(x - 1) + 0(y - 0) + 0(z - 1) = 0$$

or

$$2(x - 1) = 0.$$

The tangent plane is given by the equation

$$x = 1.$$

Problems

41. The gradient of (a) is $2x\vec{i} + 2y\vec{j} + 2z\vec{k}$, which points radially outward from the origin, so (a) goes with (III).
The gradient of (c) is parallel to the gradient of (a) but pointing inward, so (c) goes with (IV).
The gradient of (b) is $2x\vec{i} + 2y\vec{j}$, which points radially outward from the z-axis, so (b) goes with (I).
The gradient of (d) is parallel to the gradient of (b) but pointing inward, so (d) goes with (II).

45. The surface is given by $F(x, y, z) = 0$ where $F(x, y, z) = x - y^3 z^7$. The normal direction is

$$\text{grad } F = \frac{\partial F}{\partial x}\vec{i} + \frac{\partial F}{\partial y}\vec{j} + \frac{\partial F}{\partial z}\vec{k} = \vec{i} - 3y^2 z^7 \vec{j} - 7y^3 z^6 \vec{k}.$$

Thus, at $(1, -1, -1)$ a normal vector is $\vec{i} + 3\vec{j} + 7\vec{k}$. The tangent plane has the equation

$$1(x - 1) + 3(y - (-1)) + 7(z - (-1)) = 0$$
$$x + 3y + 7z = -9.$$

49. (a) A normal to the surface is given by the gradient of the function $f(x, y, z) = x^2 + y^2 + 3z^2$,

$$\text{grad } f = 2x\vec{i} + 2y\vec{j} + 6z\vec{k}.$$

At the point $(0.6, 0.8, 1)$, a normal to the surface and to the tangent plane is

$$\vec{n} = 1.2\vec{i} + 1.6\vec{i} + 6\vec{k}.$$

Since the plane goes through the point $(0.6, 0.8, 1)$, its equation is

$$1.2(x - 0.6) + 1.6(y - 0.8) + 6(z - 1) = 0$$
$$1.2x + 1.6y + 6z = 8.$$

(b) We want to know if there are x, y, z values such that a normal to the surface is parallel to the normal to the plane. That is, is $2x\vec{i} + 2y\vec{j} + 6z\vec{k}$ parallel to $8\vec{i} + 6\vec{j} + 30\vec{k}$? Yes, if $2x = 8t$, $2y = 6t$, $6z = 30t$ for some value of t. That is, if

$$x = 4t, \quad y = 3t, \quad z = 5t.$$

Substituting these equations into the equation for the surface and solving for t, we get

$$(4t)^2 + (3t)^2 + 3(5t)^2 = 4$$
$$100t^2 = 4$$
$$t = \pm\sqrt{\frac{4}{100}} = \pm 0.2.$$

So there are two points on the surface at which the tangent plane is parallel to the plane $8x + 6y + 30z = 1$. They are

$$\pm(0.8, 0.6, 1).$$

53. (a) In the direction of grad F:

$$\text{grad } F \bigg|_{(-1,1,1)} = \left((2x + 2xz^2)\vec{i} + (4y^3)\vec{j} + (2x^2 z)\vec{k}\right)\bigg|_{(-1,1,1)} = -4\vec{i} + 4\vec{j} + 2\vec{k}.$$

(b) The rate of change in the direction of grad F with respect to distance $= \|\nabla F\| = \sqrt{16 + 16 + 4} = 6$. Now we want rate of change with respect to time. If we move at 4 units/sec:

$$\text{Rate of change of } \frac{\text{Conc}}{\text{Time}} = \text{Rate of change of } \frac{\text{Conc}}{\text{Dist}} \times \text{Rate of change of } \frac{\text{Dist}}{\text{Time}}$$
$$= 6 \times 4 = 24 \text{mg/cm}^3\text{/sec.}$$

57. (a) We have $\nabla G = (2x - 5y)\vec{i} + (-5x + 2yz)\vec{j} + (y^2)\vec{k}$, so $\nabla G(1, 2, 3) = -8\vec{i} + 7\vec{j} + 4\vec{k}$. The rate of change is given by the directional derivative in the direction $\vec{v}$:

$$\text{Rate of change in density} = \nabla G \cdot \frac{\vec{v}}{\|\vec{v}\|} = (-8\vec{i} + 7\vec{j} + 4\vec{k}) \cdot \frac{(2\vec{i} + \vec{j} - 4\vec{k})}{\sqrt{21}}$$
$$= \frac{-16 + 7 - 16}{\sqrt{21}} = \frac{-25}{\sqrt{21}} \approx -5.455.$$

(b) The direction of maximum rate of change is $\nabla G(1, 2, 3) = -8\vec{i} + 7\vec{j} + 4\vec{k}$.

(c) The maximum rate of change is $\|\nabla G(1, 2, 3)\| = \sqrt{(-8)^2 + 7^2 + 4^2} = \sqrt{129} \approx 11.36$.

61. (a) is (V) since $\vec{r} + \vec{a}$ is a vector not a scalar.
  (b) is (IV) since $\text{grad}(\vec{r} \cdot \vec{a}) = \text{grad}(a_1 x + a_2 y + a_3 z) = \vec{a}$.
  (c) is (V) since $\vec{r} \times \vec{a}$ is a vector not a scalar.

65.

$$\text{grad}(\vec{\mu} \cdot \vec{r}) = \text{grad}(\mu_1 x + \mu_2 y + \mu_3 z)$$
$$= \mu_1 \vec{i} + \mu_2 \vec{j} + \mu_3 \vec{k} = \vec{\mu}.$$

Solutions for Section 14.6

Exercises

1. Using the chain rule we see:

$$\frac{dz}{dt} = \frac{\partial z}{\partial x}\frac{dx}{dt} + \frac{\partial z}{\partial y}\frac{dy}{dt}$$
$$= -y^2 e^{-t} + 2xy \cos t$$
$$= -(\sin t)^2 e^{-t} + 2e^{-t} \sin t \cos t$$
$$= \sin(t)e^{-t}(2\cos t - \sin t)$$

We can also solve the problem using one variable methods:

$$z = e^{-t}(\sin t)^2$$
$$\frac{dz}{dt} = \frac{d}{dt}(e^{-t}(\sin t)^2)$$
$$= \frac{de^{-t}}{dt}(\sin t)^2 + e^{-t}\frac{d(\sin t)^2}{dt}$$
$$= -e^{-t}(\sin t)^2 + 2e^{-t}\sin t \cos t$$
$$= e^{-t}\sin t(2\cos t - \sin t)$$

5. Using the chain rule we see:

$$\frac{dz}{dt} = \frac{\partial z}{\partial x}\frac{dx}{dt} + \frac{\partial z}{\partial y}\frac{dy}{dt}$$
$$= 2t(\sin y + y \cos x) + \frac{1}{t}(x \cos y + \sin x)$$
$$= 2t \sin(\ln t) + 2t \ln(t)\cos(t^2) + t \cos(\ln t) + \frac{\sin t^2}{t}$$

This problem can also be solved using one variable methods. Attempting to solve the problem that way will demonstrate the advantage of using the chain rule.

9. Since z is a function of two variables x and y which are functions of two variables u and v, the two chain rule identities which apply are:

$$\frac{\partial z}{\partial u} = \frac{\partial z}{\partial x}\frac{\partial x}{\partial u} + \frac{\partial z}{\partial y}\frac{\partial y}{\partial u} = e^y(2u) + xe^y(2u)$$
$$= 2ue^y(1 + x) = 2ue^{(u^2 - v^2)}(1 + u^2 + v^2).$$
$$\frac{\partial z}{\partial v} = \frac{\partial z}{\partial x}\frac{\partial x}{\partial v} + \frac{\partial z}{\partial y}\frac{\partial y}{\partial v} = e^y(2v) + xe^y(-2v)$$
$$= 2ve^y(1 - x) = 2ve^{(u^2 - v^2)}(1 - u^2 - v^2).$$

13. Since z is a function of two variables x and y which are functions of two variables u and v, the two chain rule identities which apply are:

$$\frac{\partial z}{\partial u} = \frac{\partial z}{\partial x}\frac{\partial x}{\partial u} + \frac{\partial z}{\partial y}\frac{\partial y}{\partial u}$$

$$\frac{\partial z}{\partial v} = \frac{\partial z}{\partial x}\frac{\partial x}{\partial v} + \frac{\partial z}{\partial y}\frac{\partial y}{\partial v}$$

First to find $\partial z/\partial u$

$$\frac{\partial z}{\partial u} = (e^{-y} - ye^{-x})\sin v + (-xe^{-y} + e^{-x})(-v\sin u)$$

$$= (e^{-v\cos u} - v(\cos u)e^{-u\sin v})\sin v - (-u(\sin v)e^{-v\cos u} + e^{-u\sin v})v\sin u$$

Now we find $\partial z/\partial v$ using the same method.

$$\frac{\partial z}{\partial v} = (e^{-y} - ye^{-x})u\cos v + (-xe^{-y} + e^{-x})\cos u$$

$$= (e^{-v\cos u} - v(\cos u)e^{-u\sin v})u\cos v + (-u(\sin v)e^{-v\cos u} + e^{-u\sin v})\cos u$$

Problems

17.

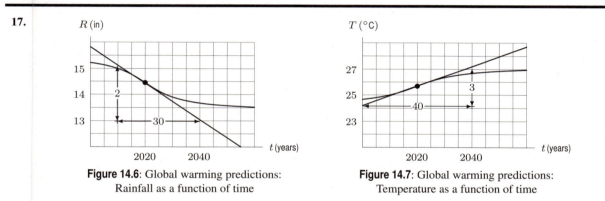

Figure 14.6: Global warming predictions: Rainfall as a function of time

Figure 14.7: Global warming predictions: Temperature as a function of time

We know that, as long as the temperature and rainfall stay close to their current values of $R = 15$ inches and $T = 30°$C, a change, ΔR, in rainfall and a change, ΔT, in temperature produces a change, ΔC, in corn production given by

$$\Delta C \approx 3.3\Delta R - 5\Delta T.$$

Now both R and T are functions of time t (in years), and we want to find the effect of a small change in time, Δt, on R and T. Figure 14.6 shows that the slope of the graph for R versus t is about $-2/30 \approx -0.07$ in/year when $t = 2020$. Similarly, Figure 14.7 shows the slope of the graph of T versus t is about $3/40 \approx 0.08°$C/year when $t = 2020$. Thus, around the year 2020,

$$\Delta R \approx -0.07\Delta t \quad \text{and} \quad \Delta T \approx 0.08\Delta t.$$

Substituting these into the equation for ΔC, we get

$$\Delta C \approx (3.3)(-0.07)\Delta t - (5)(0.08)\Delta t \approx -0.6\Delta t.$$

Since at present $C = 100$, corn production will decline by about 0.6 % between the years 2020 and 2021. Now $\Delta C \approx -0.6\Delta t$ tells us that when $t = 2020$,

$$\frac{\Delta C}{\Delta t} \approx -0.6, \quad \text{and therefore, that} \quad \frac{dC}{dt} \approx -0.6.$$

21.

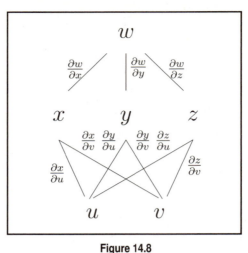

Figure 14.8

The tree diagram in Figure 14.8 tells us that

$$\frac{\partial w}{\partial u} = \frac{\partial w}{\partial x}\frac{\partial x}{\partial u} + \frac{\partial w}{\partial y}\frac{\partial y}{\partial u} + \frac{\partial w}{\partial z}\frac{\partial z}{\partial u},$$

$$\frac{\partial w}{\partial v} = \frac{\partial w}{\partial x}\frac{\partial x}{\partial v} + \frac{\partial w}{\partial y}\frac{\partial y}{\partial v} + \frac{\partial w}{\partial z}\frac{\partial z}{\partial v}.$$

25. All are done using the chain rule.

(a) We have $u = x$, $v = 3$. Thus $du/dx = 1$ and $dv/dx = 0$ so

$$f'(x) = F_u(x, 3)(1) + F_v(x, 3)(0) = F_u(x, 3).$$

(b) We have $u = 3$, $v = x$. Thus $du/dx = 0$ and $dv/dx = 1$ so

$$f'(x) = F_u(3, x)(0) + F_v(3, x)(1) = F_v(3, x).$$

(c) We have $u = x$, $v = x$. Thus $du/dx = dv/dx = 1$ so

$$f'(x) = F_u(x, x)(1) + F_v(x, x)(1) = F_u(x, x) + F_v(x, x).$$

(d) We have $u = 5x$, $v = x^2$. Thus $du/dx = 5$ and $dv/dx = 2x$ so

$$f'(x) = F_u(5x, x^2)(5) + F_v(5x, x^2)(2x).$$

29. We have $z = h(x, y)$ where $h(x, y) = f(x)g(y)$, $x = t$, and $y = t$. The chain rule gives $dz/dt = (\partial h/\partial x)dx/dt + (\partial h/\partial y)dy/dt = \partial h/\partial x + \partial h/\partial y$. Since $\partial h/\partial x = f'(x)g(y)$ and $\partial h/\partial y = f(x)g'(y)$ we have $dz/dt = f'(x)g(y) + f(x)g'(y) = f'(t)g(t) + f(t)g'(t)$.

33. To calculate $\left(\frac{\partial U}{\partial P}\right)_T$, we think of U as a function of P and T, as in $U_1(T, P)$. Thus

$$\left(\frac{\partial U}{\partial P}\right)_T = \frac{\partial U_1}{\partial P}.$$

37. (a) Thinking of V as a function of P and T gives

$$dV = \left(\frac{\partial V}{\partial P}\right)_T dP + \left(\frac{\partial V}{\partial T}\right)_P dT.$$

(b) Substituting for dV in the following expression for dU,

$$dU = \left(\frac{\partial U}{\partial P}\right)_V dP + \left(\frac{\partial U}{\partial V}\right)_P dV,$$

we get

$$dU = \left(\frac{\partial U}{\partial P}\right)_V dP + \left(\frac{\partial U}{\partial V}\right)_P \left(\left(\frac{\partial V}{\partial P}\right)_T dP + \left(\frac{\partial V}{\partial T}\right)_P dT\right).$$

Rearranging terms gives

$$dU = \left(\left(\frac{\partial U}{\partial P}\right)_V + \left(\frac{\partial U}{\partial V}\right)_P \cdot \left(\frac{\partial V}{\partial P}\right)_T\right) dP + \left(\frac{\partial U}{\partial V}\right)_P \cdot \left(\frac{\partial V}{\partial T}\right)_P dT.$$

(c) The formula for dU obtained by thinking of U as a function of P and T is

$$dU = \left(\frac{\partial U}{\partial T}\right)_P dT + \left(\frac{\partial U}{\partial P}\right)_T dP.$$

(d) Comparing coefficients of dP and dT in the two formulas gives

$$\left(\frac{\partial U}{\partial T}\right)_P = \left(\frac{\partial U}{\partial V}\right)_P \cdot \left(\frac{\partial V}{\partial T}\right)_P$$

$$\left(\frac{\partial U}{\partial P}\right)_T = \left(\frac{\partial U}{\partial P}\right)_V + \left(\frac{\partial U}{\partial V}\right)_P \cdot \left(\frac{\partial V}{\partial P}\right)_T.$$

Solutions for Section 14.7

Exercises

1. We have $f_x = 6xy + 5y^3$ and $f_y = 3x^2 + 15xy^2$, so $f_{xx} = 6y$, $f_{xy} = 6x + 15y^2$, $f_{yx} = 6x + 15y^2$, and $f_{yy} = 30xy$.

5. We have $f_x = 2ye^{2xy}$ and $f_y = 2xe^{2xy}$, so $f_{xx} = 4y^2 e^{2xy}$, $f_{xy} = 4xye^{2xy} + 2e^{2xy}$, $f_{yx} = 4xye^{2xy} + 2e^{2xy}$, and $f_{yy} = 4x^2 e^{2xy}$.

9. Since $f(x, y) = \sin(x^2 + y^2)$, we have

$$f_x = (\cos(x^2 + y^2))2x \quad , f_y = (\cos(x^2 + y^2))2y$$
$$f_{xx} = -(\sin(x^2 + y^2))4x^2 + 2\cos(x^2 + y^2)$$
$$f_{xy} = -(\sin(x^2 + y^2))4xy = f_{yx}$$
$$f_{yy} = -(\sin(x^2 + y^2))4y^2 + 2\cos(x^2 + y^2).$$

13. The quadratic Taylor expansion about $(0, 0)$ is given by

$$f(x, y) \approx Q(x, y) = f(0, 0) + f_x(0, 0)x + f_y(0, 0)y + \frac{1}{2}f_{xx}(0, 0)x^2 + f_{xy}(0, 0)xy + \frac{1}{2}f_{yy}(0, 0)y^2.$$

First we find all the relevant derivatives

$$f(x, y) = e^{-2x^2 - y^2}$$
$$f_x(x, y) = -4xe^{-2x^2 - y^2}$$
$$f_y(x, y) = -2ye^{-2x^2 - y^2}$$
$$f_{xx}(x, y) = -4e^{-2x^2 - y^2} + 16x^2 e^{-2x^2 - y^2}$$
$$f_{yy}(x, y) = -2e^{-2x^2 - y^2} + 4y^2 e^{-2x^2 - y^2}$$
$$f_{xy}(x, y) = 8xye^{-2x^2 - y^2}$$

Now we evaluate each of these derivatives at $(0, 0)$ and substitute into the formula to get as our final answer:

$$Q(x, y) = 1 - 2x^2 - y^2$$

17. The quadratic Taylor expansion about $(0, 0)$ is given by

$$f(x, y) \approx Q(x, y) = f(0, 0) + f_x(0, 0)x + f_y(0, 0)y + \frac{1}{2}f_{xx}(0, 0)x^2 + f_{xy}(0, 0)xy + \frac{1}{2}f_{yy}(0, 0)y^2.$$

So first we find all the relevant derivatives:

$$f(x, y) = \ln(1 + x^2 - y)$$
$$f_x(x, y) = \frac{2x}{1 + x^2 - y}$$
$$f_y(x, y) = \frac{-1}{1 + x^2 - y}$$
$$f_{xx}(x, y) = \frac{2(1 + x^2 - y) - 4x^2}{(1 + x^2 - y)^2}$$
$$f_{yy}(x, y) = \frac{-1}{(1 + x^2 - y)^2}$$
$$f_{xy}(x, y) = \frac{2x}{(1 + x^2 - y)^2}$$

Substituting into the formula we get as our answer:

$$Q(x, y) = -y + x^2 - \frac{y^2}{2}$$

21. (a) $f_x(P) < 0$ because f decreases as you go to the right.
(b) $f_y(P) = 0$ because f does not change as you go up.
(c) $f_{xx}(P) < 0$ because f_x decreases as you go to the right (f_x changes from a small negative number to a large negative number).
(d) $f_{yy}(P) = 0$ because f_y does not change as you go up.
(e) $f_{xy}(P) = 0$ because f_x does not change as you go up.

25. (a) $f_x(P) > 0$ because f increases as you go to the right.
(b) $f_y(P) > 0$ because f increases as you go up.
(c) $f_{xx}(P) = 0$ because f_x does not change as you go to the right. (Notice that the level curves are equidistant and parallel, so the partial derivatives of f do not change if you move horizontally or vertically.)
(d) $f_{yy}(P) = 0$ because f_y does not change as you go up.
(e) $f_{xy}(P) = 0$ because f_x does not change as you go up.

29. Since $f(x, y) = \sqrt{1 + 2x - y}$, the first and second derivatives are

$$f_x = \frac{1}{\sqrt{1 + 2x - y}}$$
$$f_y = \frac{-1}{2\sqrt{1 + 2x - y}}$$
$$f_{xx} = \frac{-1}{(1 + 2x - y)^{3/2}}$$
$$f_{xy} = \frac{1}{2(1 + 2x - y)^{3/2}}$$
$$f_{yy} = \frac{-1}{4(1 + x - 2y)^{3/2}},$$

so we find that

$$f(0, 0) = 1$$
$$f_x(0, 0) = 1$$
$$f_y(0, 0) = -1/2$$
$$f_{xx}(0, 0) = -1$$
$$f_{xy}(0, 0) = 1/2$$
$$f_{yy}(0, 0) = -1/4.$$

The best quadratic approximation for $f(x, y)$ for (x, y) near $(0, 0)$ is

$$f(x, y) \approx 1 + x - \frac{1}{2}y - \frac{1}{2}x^2 + \frac{1}{2}xy - \frac{1}{8}y^2.$$

Problems

33. We have $f(1, 0) = 1$ and the relevant derivatives are:

$$\begin{align}
f_x &= e^{-y} \quad &\text{so} \quad f_x(1, 0) &= 1 \\
f_y &= -xe^{-y} \quad &\text{so} \quad f_y(1, 0) &= -1 \\
f_{xx} &= 0 \quad &\text{so} \quad f_{xx}(1, 0) &= 0 \\
f_{xy} &= -e^{-y} \quad &\text{so} \quad f_{xy}(1, 0) &= -1 \\
f_{yy} &= xe^{-y} \quad &\text{so} \quad f_{yy}(1, 0) &= 1 \ .
\end{align}$$

Thus the linear approximation, $L(x, y)$ to $f(x, y)$ at $(1, 0)$, is given by:

$$\begin{align}
f(x, y) \approx L(x, y) &= f(1, 0) + f_x(1, 0)(x - 1) + f_y(1, 0)(y - 0) \\
&= 1 + (x - 1) - y \ .
\end{align}$$

The quadratic approximation, $Q(x, y)$ to $f(x, y)$ near $(1, 0)$, is given by:

$$\begin{align}
f(x, y) \approx Q(x, y) &= f(1, 0) + f_x(1, 0)(x - 1) + f_y(1, 0)(y - 0) + \frac{1}{2}f_{xx}(1, 0)(x - 1)^2 \\
&\quad + f_{xy}(1, 0)(x - 1)(y - 0) + \frac{1}{2}f_{yy}(1, 0)(y - 0)^2 \\
&= 1 + (x - 1) - y - (x - 1)y + \frac{1}{2}y^2 \ .
\end{align}$$

The values of the approximations are

$$\begin{align}
L(0.9, 0.2) &= 1 - 0.1 - 0.2 = 0.7 \\
Q(0.9, 0.2) &= 1 - 0.1 - 0.2 + 0.02 + 0.02 = 0.74
\end{align}$$

and the exact value is

$$f(0.9, 0.2) = (0.9)e^{-0.2} \approx 0.737$$

Observe that the quadratic approximation is closer to the exact value.

37. Differentiating, we get

$$\begin{align}
F_x &= e^x \sin y + e^y \cos x \qquad F_y = e^x \cos y + e^y \sin x \\
F_{xx} &= e^x \sin y - e^y \sin x \qquad F_{yy} = -e^x \sin y + e^y \sin x = -F_{xx}
\end{align}$$

Thus, $F_{xx} + F_{yy} = 0$.

41. Since $z_y = g(x)$, $z_{yy} = 0$, because g is a function of x only.

45. (a) Since P and Q lie on the same level curve, we have $a = k$.
   (b) We have $b = f_x$ and $c = f_y$. Since the gradient of f at P (respectively Q) points toward M or away from M, from the figure, we see $f_x(P)$ and $f_y(P)$ have opposite signs, while $f_x(Q)$ and $f_y(Q)$ have the same signs. Thus Q is the point (x_1, y_1), so P is (x_2, y_2).
   (c) Since $b = f_x(Q) > 0$ and $c = f_y(Q) > 0$, the value of f must increase as we go away from M. Thus, M must be a minimum (the surface is a valley).
   (d) Since M is a minimum, $m = f_x(P) < 0$ and $n = f_y(P) > 0$.

49. If f and g have exactly the same contour diagrams inside a circle about the origin, then any partial derivative of f of any order has the same value at the point $(0, 0)$ as the corresponding partial derivative of g. Since the Taylor polynomials of degree 2 for f and g near $(0, 0)$ are constructed from the values of the partial derivatives at $(0, 0)$, the two Taylor polynomials are identical. It makes no difference what the contour diagrams look like outside the circle. See Figure 14.9 for one possible solution.

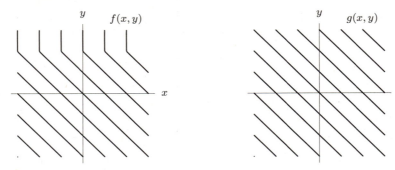

Figure 14.9

Solutions for Section 14.8

Exercises

1. Not differentiable at $(0, 0)$.

5. Not differentiable where $x = 0$; that is, not differentiable on y-axis.

9. Not differentiable at $(1, 2)$.

Problems

13. (a) The contour diagram for $f(x, y) = xy/\sqrt{x^2 + y^2}$ is shown in Figure 14.10.

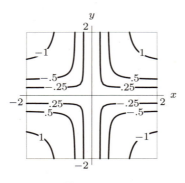

Figure 14.10

(b) By the chain rule, f is differentiable at all points (x, y) where $x^2 + y^2 \neq 0$, and so at all points $(x, y) \neq (0, 0)$.

(c) The partial derivatives of f are given by

$$f_x(x, y) = \frac{y^3}{(x^2 + y^2)^{3/2}}, \qquad \text{for} \quad (x, y) \neq (0, 0),$$

and

$$f_y(x, y) = \frac{x^3}{(x^2 + y^2)^{3/2}}, \qquad \text{for} \quad (x, y) \neq (0, 0).$$

Both f_x and f_y are continuous at $(x, y) \neq (0, 0)$.

(d) If f were differentiable at $(0,0)$, the chain rule would imply that the function

$$g(t) = \begin{cases} f(t,t), & t \neq 0 \\ 0, & t = 0 \end{cases}$$

would be differentiable at $t = 0$. But

$$g(t) = \frac{t^2}{\sqrt{2t^2}} = \frac{1}{\sqrt{2}} \cdot \frac{t^2}{|t|} = \frac{1}{\sqrt{2}} \cdot |t|,$$

which is not differentiable at $t = 0$. Hence, f is not differentiable at $(0,0)$.

(e) The partial derivatives of f at $(0,0)$ are given by

$$f_x(0,0) = \lim_{x \to 0} \frac{f(x,0) - f(0,0)}{x} = \lim_{x \to 0} \frac{\frac{x \cdot 0}{\sqrt{x^2 + 0^2}} - 0}{x} = \lim_{x \to 0} \frac{0 - 0}{x} = 0,$$

$$f_y(0,0) = \lim_{y \to 0} \frac{f(0,y) - f(0,0)}{y} = \lim_{y \to 0} \frac{\frac{0 \cdot y}{\sqrt{0^2 + y^2}} - 0}{y} = \lim_{y \to 0} \frac{0 - 0}{y} = 0.$$

The limit $\lim\limits_{(x,y) \to (0,0)} f_x(x,y)$ does not exist since if we choose $x = y = t$, $t \neq 0$, then

$$f_x(x,y) = f_x(t,t) = \frac{t^3}{(2t^2)^{3/2}} = \frac{t^3}{2\sqrt{2} \cdot |t|^3} = \begin{cases} \frac{1}{2\sqrt{2}}, & t > 0, \\ -\frac{1}{2\sqrt{2}}, & t < 0. \end{cases}$$

Thus, f_x is not continuous at $(0,0)$. Similarly, f_y is not continuous at $(0,0)$.

17. (a) The contour diagram of $f(x,y) = \sqrt{|xy|}$ is shown in Figure 14.11.

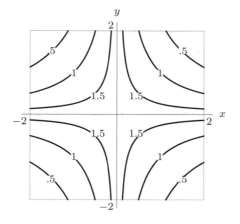

Figure 14.11

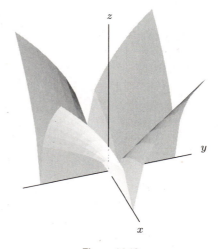

Figure 14.12

(b) The graph of $f(x,y) = \sqrt{|xy|}$ is shown in Figure 14.12.

(c) f is clearly differentiable at (x,y) where $x \neq 0$ and $y \neq 0$. So we need to look at points $(x_0, 0)$, $x_0 \neq 0$ and $(0, y_0)$, $y_0 \neq 0$. At $(x_0, 0)$:

$$f_x(x_0, 0) = \lim_{x \to x_0} \frac{f(x,0) - f(x_0, 0)}{x - x_0} = 0$$

$$f_y(x_0, 0) = \lim_{y \to 0} \frac{f(x_0, y) - f(x_0, 0)}{y} = \lim_{y \to 0} \frac{\sqrt{|x_0 y|}}{y}$$

which does not exist. So f is not differentiable at the points $(x_0, 0)$, $x_0 \neq 0$. Similarly, f is not differentiable at the points $(0, y_0)$, $y_0 \neq 0$.

(d)

$$f_x(0,0) = \lim_{x \to 0} \frac{f(x,0) - f(0,0)}{x} = 0$$

$$f_y(0,0) = \lim_{y \to 0} \frac{f(0,y) - f(0,0)}{y} = 0$$

(e) Let $\vec{u} = (\vec{i} + \vec{j})/\sqrt{2}$:

$$f_{\vec{u}}(0,0) = \lim_{t \to 0^+} \frac{f(\frac{t}{\sqrt{2}}, \frac{t}{\sqrt{2}}) - f(0,0)}{t} = \lim_{t \to 0^+} \frac{\sqrt{\frac{t^2}{2}}}{t} = \frac{1}{\sqrt{2}}.$$

We know that $\nabla f(0,0) = \vec{0}$ because both partial derivatives are 0. But if f were differentiable, $f_{\vec{u}}(0,0) = \nabla f(0,0) \cdot \vec{u} = f_x(0,0) \cdot \frac{1}{\sqrt{2}} + f_y(0,0) \cdot \frac{1}{\sqrt{2}} = 0$. But since, in fact, $f_{\vec{u}}(0,0) = 1/\sqrt{2}$, we conclude that f is not differentiable.

Solutions for Chapter 14 Review

Exercises

1. Vector. Taking the gradient and substituting $x = 1$, $y = 2$ gives

$$\operatorname{grad}(x^3 e^{-y/2})\bigg|_{(1,2)} = \left(3x^2 e^{-y/2}\vec{i} - \frac{x^3}{2}e^{-y/2}\vec{j}\right)\bigg|_{(1,2)} = 3e^{-1}\vec{i} - \frac{1}{2}e^{-1}\vec{j}.$$

5. Taking the derivative of f with respect to x and treating y as a constant we get $f_x = 2xy + 3x^2 - 7y^6$. To find f_y we take the derivative of f with respect to y and treat x as a constant. Thus, $f_y = x^2 - 42xy^5$.

9. For both partial derivatives we use the quotient rule. Thus,

$$f_x = \frac{2xy(x^2 + y^2) - x^2 y 2x}{(x^2 + y^2)^2} = \frac{2x^3 y + 2xy^3 - 2x^3 y}{(x^2 + y^2)^2} = \frac{2xy^3}{(x^2 + y^2)^2},$$

and

$$f_y = \frac{x^2(x^2 + y^2) - x^2 y 2y}{(x^2 + y^2)^2} = \frac{x^4 + x^2 y^2 - 2x^2 y^2}{(x^2 + y^2)^2} = \frac{x^4 - x^2 y^2}{(x^2 + y^2)^2}.$$

13. Since we take the derivative with respect to N, we use the power rule to obtain $f_N = c\alpha N^{\alpha - 1}V^\beta$.

17. $z_x = \dfrac{1}{2ay}(-2)\dfrac{1}{x^3} + \dfrac{15x^4 abc}{y} = -\dfrac{1}{ax^3 y} + \dfrac{15abcx^4}{y} = \dfrac{15a^2 bcx^7 - 1}{ax^3 y}$

21. Using the chain rule and the quotient rule,

$$\frac{\partial}{\partial w}\left(\frac{x^2 yw - xy^3 w^7}{w - 1}\right)^{-7/2}$$

$$= -\frac{7}{2}\left(\frac{x^2 yw - xy^3 w^7}{w - 1}\right)^{-9/2}\left(\frac{(w - 1)(x^2 y - 7xy^3 w^6) - (x^2 yw - xy^3 w^7)(1)}{(w - 1)^2}\right)$$

$$= -\frac{7}{2}\left(\frac{w - 1}{x^2 yw - xy^3 w^7}\right)^{-9/2}\left(\frac{(w - 1)(x^2 y - 7xy^3 w^6) - (x^2 yw - xy^3 w^7)}{(w - 1)^2}\right)$$

$$= \frac{7}{2}\left(\frac{w - 1}{x^2 yw - xy^3 w^7}\right)^{-9/2} \cdot \frac{x^2 y + 6xy^3 w^7 - 7xy^3 w^6}{(w - 1)^2}$$

25. The first order partial derivatives are

$$u_x = e^x \sin y, \quad u_y = e^x \cos y.$$

Thus the second order partials are

$$u_{xx} = e^x \sin y, \quad u_{yy} = -e^x \sin y.$$

29. Since $f_x = 2x$, $f_y = 2y + 3y^2$ and $f_z = 0$, we have

$$\text{grad } f = 2x\vec{i} + (2y + 3y^2)\vec{j}.$$

33. Since $f_x = 2x \sin(x^2 + y^2 + z^2)$, $f_y = 2y \sin(x^2 + y^2 + z^2)$ and $f_z = 2z \sin(x^2 + y^2 + z^2)$, we have

$$\text{grad } f = 2x \sin(x^2 + y^2 + z^2)\vec{i} + 2y \sin(x^2 + y^2 + z^2)\vec{j} + 2z \sin(x^2 + y^2 + z^2)\vec{k}.$$

37. Since the partial derivatives are

$$\frac{\partial f}{\partial x} = \cos(xy) \cdot (y) - \sin(xy) \cdot (y) = y[\cos(xy) - \sin(xy)]$$
$$\frac{\partial f}{\partial y} = \cos(xy) \cdot (x) - \sin(xy) \cdot (x) = x[\cos(xy) - \sin(xy)]$$

we have

$$\text{grad } f = \frac{\partial f}{\partial x}\vec{i} + \frac{\partial f}{\partial y}\vec{j}$$
$$= y[\cos(xy) - \sin(xy)]\vec{i} + x[\cos(xy) - \sin(xy)]\vec{j}.$$

41. We have $\text{grad } f = e^y\vec{i} + xe^y\vec{j}$, so $\text{grad } f(3, 0) = \vec{i} + 3\vec{j}$. A unit vector in the direction we want is $\vec{u} = (1/5)(4\vec{i} - 3\vec{j})$. Therefore, the directional derivative is

$$\text{grad } f(3, 0) \cdot \vec{u} = \frac{1 \cdot 4 + 3(-3)}{5} = -1.$$

45. We have $\text{grad } f = (e^{x+z} \cos y)\vec{i} - (e^{x+z} \sin y)\vec{j} + (e^{x+z} \cos y)\vec{k}$, so $\text{grad } f(1, 0, -1) = \vec{i} + \vec{k}$. A unit vector in the direction we want is $\vec{u} = (1/\sqrt{3})(\vec{i} + \vec{j} + \vec{k})$. Therefore, the directional derivative is

$$\text{grad } f(1, 0, -1) \cdot \vec{u} = \frac{1 \cdot 1 + 1 \cdot 1}{\sqrt{3}} = \frac{2}{\sqrt{3}}.$$

49. Let $f(x, y, z) = z^2 - 4x^2 - 3y^2$ so that the surface (a hyperboloid) is the level surface $f(x, y, x) = 9$. Since

$$\text{grad } f = -8x\vec{i} - 6y\vec{j} + 2z\vec{k}$$

we have

$$\text{grad } f(1, 1, 4) = -8\vec{i} - 6\vec{j} + 8\vec{k}.$$

Thus an equation of the tangent plane at $(1, 1, 4)$ is

$$-8(x - 1) - 6(y + 6) + 8(z - 4) = 0$$

so

$$-4x - 3y + 4z = 9.$$

53. The quadratic Taylor expansion about $(0,0)$ is given by

$$f(x,y) \approx Q(x,y) = f(0,0) + f_x(0,0)x + f_y(0,0)y + \frac{1}{2}f_{xx}(0,0)x^2 + f_{xy}(0,0)xy + \frac{1}{2}f_{yy}(0,0)y^2.$$

First we find all the relevant derivatives

$$\begin{aligned}
f(x,y) &= (x+1)^3(y+2) \\
f_x(x,y) &= 3(x+1)^2(y+2) \\
f_y(x,y) &= (x+1)^3 \\
f_{xx}(x,y) &= 6(x+1)(y+2) \\
f_{yy}(x,y) &= 0 \\
f_{xy}(x,y) &= 3(x+1)^2
\end{aligned}$$

Now we evaluate each of these derivatives at $(0,0)$ and substitute into the formula to get as our final answer:

$$Q(x,y) = 2 + 6x + y + 6x^2 + 3xy$$

Notice this is the same as what you get if you expand $(x+1)^3(y+2)$ and keep only the terms of degree 2 or less.

Problems

57. (a) Let $f(x,y,z) = 2x^2 - 2xy^2 + az$ so that the given surface is the level surface $f(x,y,z) = a$. Since

$$\operatorname{grad} f = (4x - 2y^2)\vec{i} - 4xy\vec{j} + a\vec{k}$$

we have

$$\operatorname{grad} f(1,1,1) = 2\vec{i} - 4\vec{j} + a\vec{k}.$$

Thus an equation of the tangent plane at $(1,1,1)$ is

$$2(x-1) - 4(y-1) + a(z-1) = 0$$

or

$$2x - 4y + az = a - 2.$$

(b) Substituting $x = 0$, $y = 0$, $z = 0$ into the equation for the tangent plane in part (a) we have $0 = a - 2$. The tangent plane passes through the origin if $a = 2$.

61. The partial derivative, $\partial Q/\partial b$ is the rate of change of the quantity of beef purchased with respect to the price of beef, when the price of chicken stays constant. If the price of beef increases and the price of chicken stays the same, we expect consumers to buy less beef and more chicken. Thus when b increases, we expect Q to decrease, so $\partial Q/\partial b < 0$.

On the other hand, $\partial Q/\partial c$ is the rate of change of the quantity of beef purchased with respect to the price of chicken, when the price of beef stays constant. An increase in the price of chicken is likely to cause consumers to buy less chicken and more beef. Thus when c increases, we expect Q to increase, so $\partial Q/\partial c > 0$.

65. We need the partial derivatives, $f_x(1,0)$ and $f_y(1,0)$. We have

$$\begin{aligned}
f_x(x,y) &= 2xe^{xy} + x^2ye^{xy}, &\quad \text{so } f_x(1,0) &= 2 \\
f_y(x,y) &= x^3e^{xy}, &\quad \text{so } f_y(1,0) &= 1.
\end{aligned}$$

(a) Since $f(1,0) = 1$, the tangent plane is

$$z = f(1,0) + 2(x-1) + 1(y-0) = 1 + 2(x-1) + y.$$

(b) The linear approximation can be obtained from the equation of the tangent plane:

$$f(x,y) \approx 1 + 2(x-1) + y.$$

(c) At $(1,0)$, the differential is

$$df = f_x dx + f_y dy = 2dx + dy.$$

69. The directional derivative is approximately the change in z (as we move in direction $\vec{v}$) divided by the horizontal change in position. The directional derivative is $f_{\vec{i}} \approx \dfrac{2-1}{4-1} = \dfrac{1}{3} \approx 0.3$.

73. The directional derivative is approximately the change in z (as we move in direction $\vec{v}$) divided by the horizontal change in position. In the direction of $\vec{v}$ we go from point $(3,3)$ to point $(1,4)$. We have the directional derivative

$$\approx \frac{2-3}{\|-2\vec{i} + \vec{j}\|} = \frac{-1}{\sqrt{5}} \approx -0.4.$$

77. (a) Incorrect. $\|\operatorname{grad} H\|$ is not the rate of change of H. In fact, there's no such thing as the rate of change of H, although directional derivatives can give its rate of change in a particular direction. For example, this expression would give the wrong answer if the ant was crawling along a contour of H, since then the rate of change of the temperature it experiences is zero even though $\|\operatorname{grad} H\|$ and $\vec{v}$ might not be zero.

(b) Correct. If $\vec{u} = \vec{v}/\|\vec{v}\|$, then

$$\operatorname{grad} H \cdot \vec{v} = \operatorname{grad} H \cdot \frac{\vec{v}}{\|\vec{v}\|}\|\vec{v}\| = (\operatorname{grad} H \cdot \vec{u})\|\vec{v}\| = H_{\vec{u}}\|\vec{v}\|$$

$$= (\text{Rate of change of } H \text{ in direction } \vec{u} \text{ in deg/cm})(\text{Speed of ant in cm/sec})$$

$$= \text{Rate of change of } H \text{ in deg/sec.}$$

(c) Incorrect, this is the directional derivative, which gives the rate of change with respect to distance, not time.

81. We have

$$\frac{\partial P}{\partial x} = 5e^{-0.1\sqrt{x^2+y^2+z^2}}(-0.1)(2x)\frac{1}{2}(x^2+y^2+z^2)^{-1/2} = -0.5e^{-0.1\sqrt{x^2+y^2+z^2}}\frac{x}{\sqrt{x^2+y^2+z^2}},$$

and similarly

$$\frac{\partial P}{\partial y} = -0.5e^{-0.1\sqrt{x^2+y^2+z^2}}\frac{y}{\sqrt{x^2+y^2+z^2}}, \quad \text{and} \quad \frac{\partial P}{\partial z} = -0.5e^{-0.1\sqrt{x^2+y^2+z^2}}\frac{z}{\sqrt{x^2+y^2+z^2}}.$$

Thus the gradient of P at $(0,0,1)$ is

$$\operatorname{grad} P = \left.\frac{\partial P}{\partial x}\right|_{(x,y,z)=(0,0,1)}\vec{i} + \left.\frac{\partial P}{\partial y}\right|_{(x,y,z)=(0,0,1)}\vec{j} + \left.\frac{\partial P}{\partial z}\right|_{(x,y,z)=(0,0,1)}\vec{k} = -0.5e^{-0.1}\vec{k}.$$

Let $\vec{u}$ be a unit vector in the direction of $\vec{v}$, so

$$\vec{u} = \frac{\vec{v}}{\|\vec{v}\|}.$$

Then

Rate of change of pressure in atm/sec = (Directional derivative in direction $\vec{u}$ in atm/mi)(Speed of spacecraft in mi/sec)

$$= P_{\vec{u}}\|\vec{v}\| = (\operatorname{grad} P \cdot \vec{u})\|\vec{v}\| = \operatorname{grad} P \cdot \frac{\vec{v}}{\|\vec{v}\|}\|\vec{v}\| = \operatorname{grad} P \cdot \vec{v}$$

$$= -0.5e^{-0.1}\vec{k} \cdot (\vec{i} - 2.5\vec{k}) = 0.5 \cdot 2.5e^{-0.1} = 1.131 \text{ atm/sec.}$$

85. (a) The chain rule gives

$$\frac{\partial z}{\partial x} = \frac{\partial z}{\partial u}\frac{\partial u}{\partial x} + \frac{\partial z}{\partial v}\frac{\partial v}{\partial x} = (2u - e^v)1 + (-ue^v)2.$$

At $(x,y) = (1,2)$, we have $u = 1 + 2 \cdot 2 = 5$ and $v = 2 \cdot 1 - 2 = 0$, so

$$\left.\frac{\partial z}{\partial y}\right|_{(x,y)=(1,2)} = (2 \cdot 5 - e^0)1 - 5e^0 \cdot 2 = -1.$$

(b) The chain rule gives

$$\frac{\partial z}{\partial y} = \frac{\partial z}{\partial u}\frac{\partial u}{\partial y} + \frac{\partial z}{\partial v}\frac{\partial v}{\partial y} = (2u - e^v)2 + (-ue^v)(-1).$$

At $(x,y) = (1,2)$, we have $(u,v) = (5,0)$, so

$$\left.\frac{\partial z}{\partial y}\right|_{(x,y)=(1,2)} = (2 \cdot 5 - e^0)2 + 5e^0 = 23.$$

89. The error, dT, in the period T is given by

$$dT = \frac{\partial T}{\partial l} dl + \frac{\partial T}{\partial g} dg,$$

where

$$\frac{\partial T}{\partial l} = \frac{\sqrt{\frac{l}{g}}\,\pi}{l}$$

and

$$\frac{\partial T}{\partial g} = -\frac{\sqrt{\frac{l}{g}}\,\pi}{g},$$

so that

$$T_l(2, 9.8) = 0.7096, \quad T_g(2, 9.8) = -0.1448.$$

We also have that

$$dl = -0.01, \quad dg = 0.01.$$

The maximum discrepancy in the period is then given by

$$dT = 0.7096(-0.01) - 0.1448(0.01) = -0.008544.$$

93. The directional derivative, or slope, of f at $(2, 1)$ in the direction perpendicular to these contours is the length of the gradient. At the point $(2, 1)$, we have grad $f = -3\vec{i} + 4\vec{j}$, so $\| \operatorname{grad} f \| = 5$. Thus if d is the distance between the contours, we have

$$\frac{7.3 - 7}{d} = \text{Slope in direction perpendicular to contours} = 5.$$

Thus $d = 0.3/5 = 0.06$.

97. By the chain rule,

$$f_r(2, 1) = f_x(2, 1) \cos \theta + f_y(2, 1) \sin \theta$$
$$f_\theta(2, 1) = f_x(2, 1)(-r \sin \theta) + f_y(2, 1)(r \cos \theta).$$

Since $x^2 + y^2 = r^2$, we have $r = \sqrt{2^2 + 1^2} = \sqrt{5}$, and $\cos \theta = x/r = 2/\sqrt{5}$, and $\sin \theta = y/r = 1/\sqrt{5}$. Thus

$$f_r(2, 1) = (-3)\frac{2}{\sqrt{5}} + (4)\frac{1}{\sqrt{5}} = -\frac{2}{\sqrt{5}}$$
$$f_\theta(2, 1) = (-3)(-1) + (4)(2) = 11.$$

For $\vec{u} = (1/\sqrt{5})(2\vec{i} + \vec{j})$, we have $f_{\vec{u}}(2, 1) = -2/\sqrt{5}$. Since $\vec{u}$ is pointing radially out from the origin to the point $(2, 1)$, we expect that $f_{\vec{u}}(2, 1) = f_r(2, 1)$.

101. (a) The first-order Taylor polynomial of a function f about a point (a, b) is equal to

$$f(a, b) + f_x(a, b)(x - a) + f_y(a, b)(y - b).$$

Computing the partial derivatives, we get:

$$f_x = 2(x - 1)e^{(x-1)^2 + (y-3)^2}$$
$$f_y = 2(y - 3)e^{(x-1)^2 + (y-3)^2}$$
$$f_x(0, 0) = 2(-1)e^{(-1)^2 + (-3)^2}$$
$$= -2e^{10}$$
$$f_y(0, 0) = 2(-3)e^{(-1)^2 + (-3)^2}$$
$$= -6e^{10}$$

Thus,

$$f(x, y) \approx e^{10} - 2e^{10}x - 6e^{10}y$$

(b) The second-order Taylor polynomial of a function f about the point $(1, 3)$ is given by

$$f(1,3) + f_x(1,3)(x-1) + f_y(1,3)(y-3)$$
$$+ \frac{1}{2}f_{xx}(1,3)(x-1)^2 + f_{xy}(1,3)(x-1)(y-3) + \frac{1}{2}f_{yy}(1,3)(y-3)^2.$$

Computing the partial derivatives, we get:

$$f_x = 2(x-1)e^{(x-1)^2+(y-3)^2}$$
$$f_y = 2(y-3)e^{(x-1)^2+(y-3)^2}$$
$$f_{xx} = (4(x-1)^2+2)e^{(x-1)^2+(y-3)^2}$$
$$f_{xy} = 4(x-1)(y-3)e^{(x-1)^2+(y-3)^2}$$
$$f_{yy} = (4(y-3)^2+2)e^{(x-1)^2+(y-3)^2}$$

Substituting in the point $(1, 3)$ to these partial derivatives, we get:

$$f_x(1,3) = 0$$
$$f_y(1,3) = 0$$
$$f_{xy}(1,3) = 0$$
$$f_{xx}(1,3) = (4(0)^2+2)e^{0^2+0^2} = 2$$
$$f_{yy}(1,3) = (4(0)^2+2)e^{0^2+0^2} = 2$$

Thus,

$$f(x,y) \approx e^0 + 0(x-1) + 0(y-3) + \frac{2}{2}(x-2)^2 + 0(x-1)(y-3) + \frac{2}{2}(y-3)^2$$
$$f(x,y) \approx 1 + (x-1)^2 + (y-3)^2.$$

(c) A vector perpendicular to the level curve is grad f. At the point $(0, 0)$, we have

$$\operatorname{grad} f = f_x(0,0)\vec{i} + f_y(0,0)\vec{j}$$

Computing partial derivatives, we have

$$f_x = 2(x-1)e^{(x-1)^2+(y-3)^2}$$
$$f_y = 2(y-3)e^{(x-1)^2+(y-3)^2}$$
$$f_x(0,0) = 2(-1)e^{(-1)^2+(-3)^2}$$
$$= -2e^{10}$$
$$f_y(0,0) = 2(-3)e^{(-1)^2+(-3)^2}$$
$$= -6e^{10}$$

Therefore, a perpendicular vector is grad $f = -2e^{10}\vec{i} - 6e^{10}\vec{j}$. Any multiple of grad f, say $-2\vec{i} - 6\vec{j}$, will do.

(d) Since the surface can be represented by the level surface

$$F(x,y,z) = f(x,y) - z = 0,$$

a vector perpendicular to the surface at $(0, 0)$ is given by

$$\operatorname{grad} F = f_x(0,0)\vec{i} + f_y(0,0)\vec{j} - \vec{k} = -2e^{10}\vec{i} - 6e^{10}\vec{j} - \vec{k}$$

CAS Challenge Problems

105. (a) We have $f(1,2) = A_0 + A_1 + 2A_2 + A_3 + 2A_4 + 4A_5$, $f_x(1,2) = A_1 + 2A_3 + 2A_4$, and $f_y(1,2) = A_2 + A_4 + 4A_5$, so the linear approximation is

$$L(x,y) = A_0 + A_1 + 2A_2 + A_3 + 2A_4 + 4A_5 + (A_1 + 2A_3 + 2A_4)(x-1) + (A_2 + A_4 + 4A_5)(y-2).$$

Also, $m(t) = 1 + B_1t$ and $n(t) = 2 + C_1t$.

(b) Using a CAS to compute the derivatives, we find that they are both the same:

$$\frac{d}{dt}f(x(t),y(t))|_{t=0} = \frac{d}{dt}l(m(t),n(t))|_{t=0} = A_1B_1 + 2A_3B_1 + 2A_4B_1 + A_2C_1 + A_4C_1 + 4A_5C_1.$$

This can be explained using the chain rule and the fact that the derivative of a function at a point is the same as the derivative of its linear approximation there:

$$\frac{d}{dt}f(x(t),y(t))|_{t=0} = \frac{\partial}{\partial x}f(x(0),y(0))x'(0) + \frac{\partial}{\partial y}f(x(0),y(0))y'(0)$$

$$= \frac{\partial}{\partial x}l(x(0),y(0))m'(0) + \frac{\partial}{\partial y}l(x(0),y(0))n'(0) = \frac{d}{dt}l(m(t),n(t))|_{t=0}$$

CHECK YOUR UNDERSTANDING

1. True. This is the instantaneous rate of change of f in the x-direction at the point $(10, 20)$.

5. True. The property $f_x(a,b) > 0$ means that f increases in the positive x-direction near (a,b), so f must decrease in the negative x-direction near (a,b).

9. True. A function with constant $f_x(x,y)$ and $f_y(x,y)$ has constant x-slope and constant y-slope, and therefore has a graph which is a plane.

13. True. Since g is a function of x only, it can be treated like a constant when taking the y partial derivative.

17. False. The equation $z = 2 + 2x(x-1) + 3y^2(y-1)$ is not linear. The correct equation is $z = 2 + 2(x-1) + 3(y-1)$, which is obtained by evaluating the partial derivatives at the point $(1,1)$.

21. False. The graph of f is a paraboloid, opening upward. The tangent plane to this surface at any point lies completely under the surface (except at the point of tangency). So the local linearization *underestimates* the value of f at nearby points.

25. True. If f is linear, then $f(x,y) = mx + ny + c$ for some m, n and c. So $f_x = m$ and $f_y = n$ giving $df = m\,dx + n\,dy$, which is linear in the variables dx and dy.

29. False. The gradient is *perpendicular* to the contour of f at (a,b).

33. False. The gradient vector at $(3,4)$ has no relation to the direction of the vector $3\vec{i} + 4\vec{j}$. For example, if $f(x,y) = x+2y$, then grad $f = \vec{i} + 2\vec{j}$, which is not perpendicular to $3\vec{i} + 4\vec{j}$. The gradient vector grad $f(3,4)$ *is* perpendicular to the contour of f passing through the point $(3,4)$.

37. True. The length of the gradient gives the maximal directional derivative in any direction. The gradient vector is grad $f(0,0) = \vec{i} + \vec{j}$, which has length $\sqrt{2}$.

41. True. The definition of $f_x(x,y)$ is the limit of the difference quotient

$$f_x(x,y) = \lim_{h\to 0}\frac{f(x+h,y) - f(x,y)}{h}.$$

The symmetry of f gives

$$\frac{f(x+h,y)-f(x,y)}{h} = \frac{f(y,x+h)-f(y,x)}{h}.$$

The definition of $f_y(y,x)$ is the limit of the difference quotient

$$f_y(y,x) = \lim_{h\to 0}\frac{f(y,x+h)-f(y,x)}{h}.$$

Thus $f_x(x,y) = f_y(y,x)$.

45. True. Take the direction perpendicular to grad f at that point. If grad $f = 0$, any direction will do.

CHAPTER FIFTEEN

Solutions for Section 15.1

Exercises

1. The point A is not a critical point and the contour lines look like parallel lines. The point B is a critical point and is a local maximum; the point C is a saddle point.

5. At the origin, the second derivative test gives

$$D = k_{xx}k_{yy} - (k_{xy})^2 = \left((-\sin x \sin y)(-\sin x \sin y) - (\cos x \cos y)^2\right)\Big|_{x=0, y=0}$$
$$= \sin^2 0 \sin^2 0 - \cos^2 0 \cos^2 0$$
$$= -1 < 0.$$

Thus $k(0,0)$ is a saddle point.

9. To find the critical points, we solve $f_x = 0$ and $f_y = 0$ for x and y. Solving

$$f_x = 3x^2 - 6x = 0$$
$$f_y = 2y + 10 = 0$$

shows that $x = 0$ or $x = 2$ and $y = -5$. There are two critical points: $(0, -5)$ and $(2, -5)$.

We have

$$D = (f_{xx})(f_{yy}) - (f_{xy})^2 = (6x - 6)(2) - (0)^2 = 12x - 12.$$

When $x = 0$, we have $D = -12 < 0$, so f has a saddle point at $(0, -5)$. When $x = 2$, we have $D = 12 > 0$ and $f_{xx} = 6 > 0$, so f has a local minimum at $(2, -5)$.

13. Find the critical point(s) by setting

$$f_x = (xy + 1) + (x + y) \cdot y = y^2 + 2xy + 1 = 0,$$
$$f_y = (xy + 1) + (x + y) \cdot x = x^2 + 2xy + 1 = 0,$$

then we get $x^2 = y^2$, and so $x = y$ or $x = -y$.

If $x = y$, then $x^2 + 2x^2 + 1 = 0$, that is, $3x^2 = -1$, and there is no real solution. If $x = -y$, then $x^2 - 2x^2 + 1 = 0$, which gives $x^2 = 1$. Solving it we get $x = 1$ or $x = -1$, then $y = -1$ or $y = 1$, respectively. Hence, $(1, -1)$ and $(-1, 1)$ are critical points.
Since

$$f_{xx}(x, y) = 2y,$$
$$f_{xy}(x, y) = 2y + 2x \quad \text{and}$$
$$f_{yy}(x, y) = 2x,$$

the discriminant is

$$D(x, y) = f_{xx}f_{yy} - f_{xy}^2$$
$$= 2y \cdot 2x - (2y + 2x)^2$$
$$= -4(x^2 + xy + y^2).$$

thus

$$D(1, -1) = -4(1^2 + 1 \cdot (-1) + (-1)^2) = -4 < 0,$$
$$D(-1, 1) = -4((-1)^2 + (-1) \cdot 1 + 1^2) = -4 < 0.$$

Therefore $(1, -1)$ and $(-1, 1)$ are saddle points.

Problems

17. At the critical point $x = 1, y = 0$,

$$f_x = 2x + A = 0, \quad \text{so } 2 + A = 0 \text{ or } A = -2$$
$$f_y = 2y = 0.$$

Thus, $f(x, y) = x^2 - 2x + y^2 + B$ has a critical point at $(1, 0)$. Since $f_{xx} = 2$ and $f_{yy} = 2$ and $f_{xy} = 0$ at $(1, 0)$,

$$D = f_{xx}f_{yy} - (f_{xy})^2 = 2 \cdot 2 - 0^2 = 4,$$

so the second derivative test shows the critical point at $(1, 0)$ is a local minimum. The value of the minimum is

$$f(1, 0) = 1^2 - 2 \cdot 1 + 0^2 + B = 20, \quad \text{so } B = 21.$$

21. (a) P is a local maximum.
(b) Q is a saddle point.
(c) R is a local minimum.
(d) S is none of these.

25. At a critical point,

$$f_x = \cos x \sin y = 0 \quad \text{so} \quad \cos x = 0 \text{ or } \sin y = 0;$$

and

$$f_y = \sin x \cos y = 0 \quad \text{so} \quad \sin x = 0 \text{ or } \cos y = 0.$$

Case 1: Assume $\cos x = 0$. This gives

$$x = \cdots - \frac{3\pi}{2}, -\frac{\pi}{2}, \frac{\pi}{2}, \frac{3\pi}{2}, \cdots$$

(This can be written more compactly as: $x = k\pi + \pi/2$, for $k = 0, \pm 1, \pm 2, \cdots$.)
If $\cos x = 0$, then $\sin x = \pm 1 \neq 0$. Thus in order to have $f_y = 0$ we need $\cos y = 0$, giving

$$y = \cdots - \frac{3\pi}{2}, -\frac{\pi}{2}, \frac{\pi}{2}, \frac{3\pi}{2}, \cdots$$

(More compactly, $y = l\pi + \pi/2$, for $l = 0, \pm 1, \pm 2, \cdots$)
Case 2: Assume $\sin y = 0$. This gives

$$y = \cdots - 2\pi, -\pi, 0, \pi, 2\pi, \cdots$$

(More compactly, $y = l\pi$, for $l = 0, \pm 1, \pm 2, \cdots$)
If $\sin y = 0$, then $\cos y = \pm 1 \neq 0$, so to get $f_y = 0$ we need $\sin x = 0$, giving

$$x = \cdots, -2\pi, -\pi, 0, \pi, 2\pi, \cdots$$

(More compactly, $x = k\pi$ for $k = 0, \pm 0, \pm 1, \pm 2, \cdots$)
Hence we get all the critical points of $f(x, y)$. Those from Case 1 are as follows:

$$\cdots \left(-\frac{\pi}{2}, -\frac{\pi}{2}\right), \left(-\frac{\pi}{2}, \frac{\pi}{2}\right), \left(-\frac{\pi}{2}, \frac{3\pi}{2}\right) \cdots$$

$$\cdots \left(\frac{\pi}{2}, -\frac{\pi}{2}\right), \left(\frac{\pi}{2}, \frac{\pi}{2}\right), \left(\frac{\pi}{2}, \frac{3\pi}{2}\right) \cdots$$

$$\cdots \left(\frac{3\pi}{2}, -\frac{\pi}{2}\right), \left(\frac{3\pi}{2}, \frac{\pi}{2}\right), \left(\frac{3\pi}{2}, \frac{3\pi}{2}\right) \cdots$$

Those from Case 2 are as follows:

$$\cdots (-\pi, -\pi), (-\pi, 0), (-\pi, \pi), (-\pi, 2\pi) \cdots$$

$$\cdots (0, -\pi), (0, 0), (0, \pi), (0, 2\pi) \cdots$$

$$\cdots (\pi, -\pi), (\pi, 0), (\pi, \pi), (\pi, 2\pi) \cdots$$

More compactly these points can be written as,

$$(k\pi, l\pi), \quad \text{for} \ \ k = 0, \pm 1, \pm 2, \cdots, l = 0, \pm 1, \pm 2, \cdots$$

$$\text{and} \ (k\pi + \frac{\pi}{2}, l\pi + \frac{\pi}{2}), \quad \text{for} \ \ k = 0, \pm 1, \pm 2, \cdots, l = 0, \pm 1, \pm 2, \cdots$$

To classify the critical points, we find the discriminant. We have

$$f_{xx} = -\sin x \sin y, \qquad f_{yy} = -\sin x \sin y, \qquad \text{and} \qquad f_{xy} = \cos x \cos y.$$

Thus the discriminant is

$$\begin{aligned} D(x, y) &= f_{xx}f_{yy} - f_{xy}^2 \\ &= (-\sin x \sin y)(-\sin x \sin y) - (\cos x \cos y)^2 \\ &= \sin^2 x \sin^2 y - \cos^2 x \cos^2 y \\ &= \sin^2 y - \cos^2 x. \quad (\text{Use: } \sin^2 x = 1 - \cos^2 x \ \text{and factor.}) \end{aligned}$$

At points of the form $(k\pi, l\pi)$ where $k = 0, \pm 1, \pm 2, \cdots; l = 0, \pm 1, \pm 2, \cdots$, we have
$D(x, y) = -1 < 0$ so $(k\pi, l\pi)$ are saddle points.
At points of the form $(k\pi + \frac{\pi}{2}, l\pi + \frac{\pi}{2})$ where $k = 0, \pm 1, \pm 2, \cdots; l = 0, \pm 1, \pm 2, \cdots$
$D(k\pi + \frac{\pi}{2}, l\pi + \frac{\pi}{2}) = 1 > 0$, so we have two cases:
If k and l are both even or k and l are both odd, then
$f_{xx} = -\sin x \sin y = -1 < 0$, so $(k\pi + \frac{\pi}{2}, l\pi + \frac{\pi}{2})$ are local maximum points.
If k is even but l is odd or k is odd but l is even, then
$f_{xx} = 1 > 0$ so $(k\pi + \frac{\pi}{2}, l\pi + \frac{\pi}{2})$ are local minimum points.

29. (a) (a, b) is a critical point. Since the discriminant $D = f_{xx}f_{yy} - f_{xy}^2 = -f_{xy}^2 < 0$, (a, b) is a saddle point.
(b) See Figure 15.1.

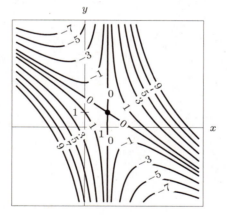

Figure 15.1

33. The first order partial derivatives are

$$f_x(x, y) = 2kx - 2y \qquad \text{and} \qquad f_y(x, y) = 2ky - 2x.$$

And the second order partial derivatives are

$$f_{xx}(x, y) = 2k \qquad f_{xy}(x, y) = -2 \qquad f_{yy}(x, y) = 2k$$

Since $f_x(0, 0) = f_y(0, 0) = 0$, the point $(0, 0)$ is a critical point. The discriminant is

$$D = (2k)(2k) - 4 = 4(k^2 - 1).$$

For $k = \pm 2$, the discriminant is positive, $D = 12$. When $k = 2$, $f_{xx}(0,0) = 4$ which is positive so we have a local minimum at the origin. When $k = -2$, $f_{xx}(0,0) = -4$ so we have a local maximum at the origin. In the case $k = 0$, $D = -4$ so the origin is a saddle point.

Lastly, when $k = \pm 1$ the discriminant is zero, so the second derivative test can tell us nothing. Luckily, we can factor $f(x, y)$ when $k = \pm 1$. When $k = 1$,

$$f(x, y) = x^2 - 2xy + y^2 = (x - y)^2.$$

This is always greater than or equal to zero. So $f(0,0) = 0$ is a minimum and the surface is a trough-shaped parabolic cylinder with its base along the line $x = y$.

When $k = -1$,

$$f(x, y) = -x^2 - 2xy - y^2 = -(x + y)^2.$$

This is always less than or equal to zero. So $f(0,0) = 0$ is a maximum. The surface is a parabolic cylinder, with its top ridge along the line $x = -y$.

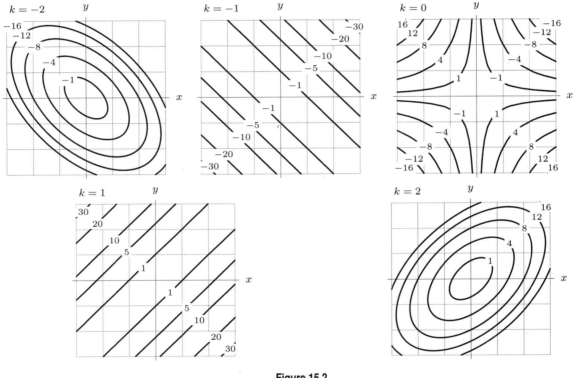

Figure 15.2

Solutions for Section 15.2

Exercises

1. Mississippi lies entirely within a region designated as 80s so we expect both the maximum and minimum daily high temperatures within the state to be in the 80s. The southwestern-most corner of the state is close to a region designated as 90s, so we would expect the temperature here to be in the high 80s, say 87-88. The northern-most portion of the state is located near the center of the 80s region. We might expect the high temperature there to be between 83-87.

Alabama also lies completely within a region designated as 80s so both the high and low daily high temperatures within the state are in the 80s. The southeastern tip of the state is close to a 90s region so we would expect the temperature here to be about 88-89 degrees. The northern-most part of the state is near the center of the 80s region so the temperature there is 83-87 degrees.

Pennsylvania is also in the 80s region, but it is touched by the boundary line between the 80s and a 70s region. Thus we expect the low daily high temperature to occur there and be about 80 degrees. The state is also touched by a boundary line of a 90s region so the high will occur there and be 89-90 degrees.

New York is split by a boundary between an 80s and a 70s region, so the northern portion of the state is likely to be about 74-76 while the southern portion is likely to be in the low 80s, maybe 81-84 or so.

California contains many different zones. The northern coastal areas will probably have the daily high as low as 65-68, although without another contour on that side, it is difficult to judge how quickly the temperature is dropping off to the west. The tip of Southern California is in a 100s region, so there we expect the daily high to be 100-101.

Arizona will have a low daily high around 85-87 in the northwest corner and a high in the 100s, perhaps 102-107 in its southern regions.

Massachusetts will probably have a high daily high around 81-84 and a low daily high of 70.

5. To maximize $z = x^2 + y^2$, it suffices to maximize x^2 and y^2. We can maximize both of these at the same time by taking the point $(1, 1)$, where $z = 2$. It occurs on the boundary of the square. (Note: We also have maxima at the points $(-1, -1), (-1, 1)$ and $(1, -1)$ which are on the boundary of the square.)

To minimize $z = x^2 + y^2$, we choose the point $(0, 0)$, where $z = 0$. It does not occur on the boundary of the square.

9. The function f has no global maximum or global minimum.

13. Suppose x is fixed. Then for large values of y the sign of f is determined by the highest power of y, namely y^3. Thus,

$$f(x, y) \to \infty \quad \text{as} \quad y \to \infty$$
$$f(x, y) \to -\infty \quad \text{as} \quad y \to -\infty.$$

So f does not have a global maximum or minimum.

Problems

17. Let the sides be x, y, z cm. Then the volume is given by $V = xyz = 32$.

The surface area S is given by

$$S = 2xy + 2xz + 2yz.$$

Substituting $z = 32/(xy)$ gives

$$S = 2xy + \frac{64}{y} + \frac{64}{x}.$$

At a critical point,

$$\frac{\partial S}{\partial x} = 2y - \frac{64}{x^2} = 0$$
$$\frac{\partial S}{\partial y} = 2x - \frac{64}{y^2} = 0,$$

The symmetry of the equations (or by dividing the equations) tells us that $x = y$ and

$$2x - \frac{64}{x^2} = 0$$
$$x^3 = 32$$
$$x = 32^{1/3} = 3.17 \text{ cm}.$$

Thus the only critical point is $x = y = (32)^{1/3}$ cm and $z = 32/\left((32)^{1/3} \cdot (32)^{1/3}\right) = (32)^{1/3}$ cm. At the critical point

$$S_{xx}S_{yy} - (S_{xy})^2 = \frac{128}{x^3} \cdot \frac{128}{y^3} - 2^2 = \frac{(128)^2}{x^3 y^3} - 4.$$

Since $D > 0$ and $S_{xx} > 0$ at this critical point, the critical point $x = y = z = (32)^{1/3}$ is a local minimum. Since $S \to \infty$ as $x, y \to \infty$, the local minimum is a global minimum.

21. The box is shown in Figure 15.3. Cost of four sides $= (2hl + 2wh)(1)¢$. Cost of two bottoms $= (2wl)(2)¢$. Thus the total cost C (in cents) of the box is

$$C = 2(hl + wh) + 4wl.$$

But volume $wlh = 512$, so $l = 512/(wh)$, thus

$$C = \frac{1024}{w} + 2wh + \frac{2048}{h}.$$

To minimize C, find the critical points of C by solving

$$C_h = 2w - \frac{2048}{h^2} = 0,$$

$$C_w = 2h - \frac{1024}{w^2} = 0.$$

We get

$$2wh^2 = 2048$$
$$2hw^2 = 1024.$$

Since $w, h \neq 0$, we can divide the first equation by the second giving

$$\frac{2wh^2}{2hw^2} = \frac{2048}{1024},$$

so

$$\frac{h}{w} = 2,$$

thus

$$h = 2w.$$

Substituting this in $C_h = 0$, we obtain $h^3 = 2048$, so $h = 12.7$ cm. Thus $w = h/2 = 6.35$ cm, and $l = 512/(wh) = 6.35$ cm. Now we check that these dimensions minimize the cost C. We find that

$$D = C_{hh}C_{ww} - C_{hw}^2 = (\frac{4096}{h^3})(\frac{2048}{w^3}) - 2^2,$$

and at $h = 12.7$, $w = 6.35$, $C_{hh} > 0$ and $D = 16 - 4 > 0$, thus C has a local minimum at $h = 12.7$ and $w = 6.35$. Since C increases without bound as $w, h \to 0$ or ∞, this local minimum must be a global minimum.

Therefore, the dimensions of the box that minimize the cost are $w = 6.35$ cm, $l = 6.35$ cm and $h = 12.7$ cm.

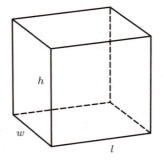

Figure 15.3

25. The total revenue is

$$R = pq = (60 - 0.04q)q = 60q - 0.04q^2,$$

and as $q = q_1 + q_2$, this gives

$$R = 60q_1 + 60q_2 - 0.04q_1^2 - 0.08q_1q_2 - 0.04q_2^2.$$

Therefore, the profit is

$$P(q_1, q_2) = R - C_1 - C_2$$
$$= -13.7 + 60q_1 + 60q_2 - 0.07q_1^2 - 0.08q_2^2 - 0.08q_1q_2.$$

At a local maximum point, we have grad $P = \vec{0}$:

$$\frac{\partial P}{\partial q_1} = 60 - 0.14q_1 - 0.08q_2 = 0,$$

$$\frac{\partial P}{\partial q_2} = 60 - 0.16q_2 - 0.08q_1 = 0.$$

Solving these equations, we find that
$$q_1 = 300 \quad \text{and} \quad q_2 = 225.$$

To see whether or not we have found a local maximum, we compute the second-order partial derivatives:
$$\frac{\partial^2 P}{\partial q_1^2} = -0.14, \quad \frac{\partial^2 P}{\partial q_2^2} = -0.16, \quad \frac{\partial^2 P}{\partial q_1 \partial q_2} = -0.08.$$

Therefore,
$$D = \frac{\partial^2 P}{\partial q_1^2}\frac{\partial^2 P}{\partial q_2^2} - \frac{\partial^2 P}{\partial q_1 \partial q_2} = (-0.14)(-0.16) - (-0.08)^2 = 0.016,$$

and so we have found a local maximum point. The graph of $P(q_1, q_2)$ has the shape of an upside down paraboloid since P is quadratic in q_1 and q_2, hence $(300, 225)$ is a global maximum point.

29. (a) We have $f(2, 1) = 120$.
  (i) If $x > 20$ then $f(x, y) > 10x > 200 > f(2, 1)$.
  (ii) If $y > 20$ then $f(x, y) > 20y > 400 > f(2, 1)$.
  (iii) If $x < 0.01$ and $y \le 20$ then $f(x, y) > 80/(xy) > 80/((0.01)(20)) = 400 > f(2, 1)$.
  (iv) If $y < 0.01$ and $x \le 20$ then $f(x, y) > 80/(xy) > 80/((20)(0.01)) = 400 > f(2, 1)$.

(b) The continuous function f must achieve a minimum at some point (x_0, y_0) in the closed and bounded region R' : $0.01 \le x \le 20, 0.01 \le y \le 20$. Since $(2, 1)$ is in R', we must have $f(x_0, y_0) \le f(2, 1)$. By part (a), $f(x_0, y_0)$ is less than all values of f in the part of R that is outside R', so $f(x_0, y_0)$ is a minimum for f on all of R. Since (x_0, y_0) is not on the boundary of R, it must be a critical point of f.

(c) The only critical point of f in R is the point $(2, 1)$, so by part (b) f has a global minimum there.

Solutions for Section 15.3

Exercises

1. Our objective function is $f(x, y) = x + y$ and our equation of constraint is $g(x, y) = x^2 + y^2 = 1$. To optimize $f(x, y)$ with Lagrange multipliers, we solve $\nabla f(x, y) = \lambda \nabla g(x, y)$ subject to $g(x, y) = 1$. The gradients of f and g are
$$\nabla f(x, y) = \vec{i} + \vec{j},$$
$$\nabla g(x, y) = 2x\vec{i} + 2y\vec{j}.$$

So the equation $\nabla f = \lambda \nabla g$ becomes
$$\vec{i} + \vec{j} = \lambda(2x\vec{i} + 2y\vec{j})$$

Solving for λ gives
$$\lambda = \frac{1}{2x} = \frac{1}{2y},$$

which tells us that $x = y$. Going back to our equation of constraint, we use the substitution $x = y$ to solve for y:
$$g(y, y) = y^2 + y^2 = 1$$
$$2y^2 = 1$$
$$y^2 = \frac{1}{2}$$
$$y = \pm\sqrt{\frac{1}{2}} = \pm\frac{\sqrt{2}}{2}.$$

Since $x = y$, our critical points are $(\frac{\sqrt{2}}{2}, \frac{\sqrt{2}}{2})$ and $(-\frac{\sqrt{2}}{2}, -\frac{\sqrt{2}}{2})$. Since the constraint is closed and bounded, maximum and minimum values of f subject to the constraint exist. Evaluating f at the critical points we find that the maximum value is $f(\frac{\sqrt{2}}{2}, \frac{\sqrt{2}}{2}) = \sqrt{2}$ and the minimum value is $f(-\frac{\sqrt{2}}{2}, -\frac{\sqrt{2}}{2}) = -\sqrt{2}$.

5. Our objective function is $f(x, y) = 3x - 2y$ and our equation of constraint is $g(x, y) = x^2 + 2y^2 = 44$. Their gradients are

$$\nabla f(x, y) = 3\vec{i} - 2\vec{j},$$
$$\nabla g(x, y) = 2x\vec{i} + 4y\vec{j}.$$

So the equation $\nabla f = \lambda \nabla g$ becomes $3\vec{i} - 2\vec{j} = \lambda(2x\vec{i} + 4y\vec{j})$. Solving for λ gives us

$$\lambda = \frac{3}{2x} = \frac{-2}{4y},$$

which we can use to find x in terms of y:

$$\frac{3}{2x} = \frac{-2}{4y}$$
$$-4x = 12y$$
$$x = -3y.$$

Using this relation in our equation of constraint, we can solve for y:

$$x^2 + 2y^2 = 44$$
$$(-3y)^2 + 2y^2 = 44$$
$$9y^2 + 2y^2 = 44$$
$$11y^2 = 44$$
$$y^2 = 4$$
$$y = \pm 2.$$

Thus, the critical points are $(-6, 2)$ and $(6, -2)$. Since the constraint is closed and bounded, maximum and minimum values of f subject to the constraint exist. Evaluating f at the critical points, we find that the maximum is $f(6, -2) = 18 + 4 = 22$ and the minimum value is $f(-6, 2) = -18 - 4 = -22$.

9. The objective function is $f(x, y, z) = x + 3y + 5z$ and the equation of constraint is $g(x, y, z) = x^2 + y^2 + z^2 = 1$. Their gradients are

$$\nabla f(x, y, z) = \vec{i} + 3\vec{j} + 5\vec{k},$$
$$\nabla g(x, y, z) = 2x\vec{i} + 2y\vec{j} + 2z\vec{k}.$$

So the equation $\nabla f = \lambda \nabla g$ becomes $\vec{i} + 3\vec{j} + 5\vec{k} = \lambda(2x\vec{i} + 2y\vec{j} + 2z\vec{k})$. Solving for λ we find

$$\lambda = \frac{1}{2x} = \frac{3}{2y} = \frac{5}{2z}.$$

Which provides us with the equations

$$2y = 6x$$
$$10x = 2z.$$

Solving the first equation for y gives us $y = 3x$. Solving the second equation for z gives us $z = 5x$. Substituting these into the equation of constraint, we can find x:

$$x^2 + (3x)^2 + (5x)^2 = 1$$
$$x^2 + 9x^2 + 25x^2 = 1$$
$$35x^2 = 1$$
$$x^2 = \frac{1}{35}$$
$$x = \pm\sqrt{\frac{1}{35}} = \pm\frac{\sqrt{35}}{35}.$$

Since $y = 3x$ and $z = 5x$, the critical points are at $\pm(\frac{\sqrt{35}}{35}, 3\frac{\sqrt{35}}{35}, \frac{\sqrt{35}}{7})$. Since the constraint is closed and bounded, maximum and minimum values of f subject to the constraint exist. Evaluating f at the critical points, we find the maximum is $f(\frac{\sqrt{35}}{35}, 3\frac{\sqrt{35}}{35}, \frac{\sqrt{35}}{7}) = \sqrt{35}\frac{35}{35} = \sqrt{35}$, and the minimum value is $f(-\frac{\sqrt{35}}{35}, -3\frac{\sqrt{35}}{35}, -\frac{\sqrt{35}}{7}) = -\sqrt{35}$.

13. The region $x^2 + y^2 \leq 2$ is the shaded disk of radius $\sqrt{2}$ centered at the origin (including the circle $x^2 + y^2 = 2$) as shown in Figure 15.4.

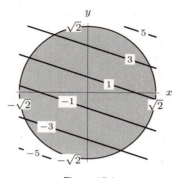

Figure 15.4

We first find the local maxima and minima of f in the interior of our disk. So we need to find the extrema of

$$f(x, y) = x + 3y, \quad \text{in the region} \quad x^2 + y^2 < 2.$$

As

$$f_x = 1$$
$$f_y = 3$$

f does not have critical points. Now let's find the local extrema of f on the boundary of the disk. We want to find the extrema of $f(x, y) = x + 3y$ subject to the constraint $g(x, y) = x^2 + y^2 - 2 = 0$. We use Lagrange multipliers

$$\text{grad } f = \lambda \, \text{grad } g \quad \text{and} \quad x^2 + y^2 = 2,$$

which give

$$1 = 2\lambda x$$
$$3 = 2\lambda y$$
$$x^2 + y^2 = 2.$$

As λ cannot be zero, we solve for x and y in the first two equations and get $x = \frac{1}{2\lambda}$ and $y = \frac{3}{2\lambda}$. Plugging into the third equation gives

$$8\lambda^2 = 10$$

so $\lambda = \pm\frac{\sqrt{5}}{2}$ and we get the solutions $(\frac{1}{\sqrt{5}}, \frac{3}{\sqrt{5}})$ and $(-\frac{1}{\sqrt{5}}, -\frac{3}{\sqrt{5}})$. Evaluating f at these points gives

$$f(\frac{1}{\sqrt{5}}, \frac{3}{\sqrt{5}}) = 2\sqrt{5} \quad \text{and}$$
$$f(-\frac{1}{\sqrt{5}}, -\frac{3}{\sqrt{5}}) = -2\sqrt{5}$$

The region $x^2 + y^2 \leq 2$ is closed and bounded, so maximum and minimum values of f in the region exist. Therefore $(\frac{1}{\sqrt{5}}, \frac{3}{\sqrt{5}})$ is a global maximum of f and $(-\frac{1}{\sqrt{5}}, -\frac{3}{\sqrt{5}})$ is a global minimum of f on the whole region $x^2 + y^2 \leq 2$.

17. We first find the critical points of f:

$$f_x = 2xy = 0, \quad f_y = x^2 + 6y - 1 = 0.$$

From the first equation, we get either $x = 0$ or $y = 0$. If $x = 0$, from the second equation we get $6y - 1 = 0$ so $y = 1/6$. If instead $y = 0$, then from the second equation $x = \pm 1$. We conclude that the critical points are $(0, 1/6)$, $(1, 0)$, and $(-1, 0)$. All three critical points satisfy the constraint $x^2 + y^2 \leq 10$.

The Lagrange conditions, grad $f = \lambda$ grad g, are:

$$2xy = \lambda 2x, \qquad x^2 + 6y - 1 = \lambda 2y$$

From the first equation, when $x \neq 0$, we divide by x to get $\lambda = y$. Substituting into the second equation, we get

$$x^2 + 6y - 1 = 2y^2.$$

Then using $x^2 = 10 - y^2$ from the constraint, we have

$$10 - y^2 + 6y - 1 = 2y^2,$$

so $3y^2 - 6y - 9 = 0$. Factoring, we get $3(y - 3)(y + 1) = 0$. From the constraint, we get $x = \pm 1$ when $y = 3$ and $x = \pm 3$ for $y = -1$. If instead $x = 0$, so that we cannot divide by x in the first Lagrange equation, then from the constraint, $y = \pm\sqrt{10}$. Summarizing, the following points are either critical points or satisfy the Lagrange conditions:

$$(1, 0), (-1, 0), (0, 1/6), (\pm 1, 3), (\pm 3, -1), (0, \pm\sqrt{10}).$$

These are the candidates for global maximum or minimum points. The corresponding values for $f(x, y) = x^2 y + 3y^2 - y$ are:

$$0, 0, -1/12, 27, -5, 30 \mp \sqrt{10}.$$

The largest value is $30 + \sqrt{10}$ at the point $(0, -\sqrt{10})$ and the smallest value is -5 at $(\pm 3, -1)$.

Problems

21. (a) C. The vectors $\vec{v}$ and $\vec{w}$ may, or may not, be parallel.
   (b) F. Since $\vec{v} \cdot \vec{w} = 2 = ||\vec{v}|| \cdot ||\vec{w}|| \cos\theta$, we must have $\cos\theta > 0$, so $0 \leq \theta < \pi/2$. Thus, the statement $\pi/2 < \theta < \pi$ is false.
   (c) T. Since $||\vec{v}|| \cdot ||\vec{w}|| \cos\theta = 2$ and $\cos\theta \leq 1$, it is true that either $||\vec{v}||$ or $||\vec{w}||$ or both must be greater than 1.
   (d) F. Since P is the hottest point on R, we know grad $g = \vec{v}$ is perpendicular to R at P.
   (e) F. Since $\vec{v}$ is perpendicular to R and $\vec{w}$ is not perpendicular to $\vec{v}$ (because $\vec{v} \cdot \vec{w} \neq 0$), we know that $\vec{w}$ is not tangent to R at P.

25. The maximum and minimum values change by approximately $\lambda\Delta c$. The Lagrange conditions give:

$$3 = \lambda 2x, \quad -2 = \lambda 4y.$$

Solving for λ and setting the expressions equal, we get $x = -3y$. Substituting into the constraint, we get $y = \pm 2$, so the points satisfying the Lagrange conditions are $(-6, 2)$ and $(6, -2)$. The corresponding values of $f(x, y) = 3x - 2y$ are -22 and 22. From the first equation, we have $\lambda = 3/(2x)$. Thus the minimum value changes by $3/(-12)\Delta c = -\Delta c/4$ and the maximum changes by $3/(12)\Delta c = \Delta c/4$.

29. (a) We want to minimize C subject to $g = x + y = 39$. Solving $\nabla C = \lambda\nabla g$ gives

$$10x + 2y = \lambda$$
$$2x + 6y = \lambda$$

so $y = 2x$. Solving with $x + y = 39$ gives $x = 13$, $y = 26$, $\lambda = 182$. Therefore $C = \$4349$.
   (b) Since $\lambda = 182$, increasing production by 1 will cause costs to increase by approximately $\$182$. (because $\lambda = \frac{||\nabla C||}{||\nabla g||} =$ rate of change of C with g). Similarly, decreasing production by 1 will save approximately $\$182$.

33. (a) Let $g(x, y) = x + y$. We are minimizing $f(x, y) = x^2 + 2y^2$ subject to the constraint $g(x, y) = c$.
   The method of Lagrange multipliers is to solve the equations

$$f_x = \lambda g_x \qquad f_y = \lambda g_y \qquad g = c,$$

which are

$$2x = \lambda \qquad 4y = \lambda \qquad x + y = c.$$

We have

$$x = \frac{2c}{3} \qquad y = \frac{c}{3} \qquad \lambda = \frac{4c}{3},$$

so there is a critical point at $(2c/3, c/3)$. Since moving away from the origin increases values of f in Figure 15.5, we see that f has a minimum on the constraint. The minimum value is

$$m(c) = f\left(\frac{2c}{3}, \frac{c}{3}\right) = \frac{2c^2}{3}.$$

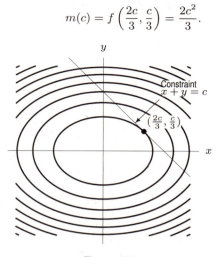

Figure 15.5

(b) Calculations in part (a) showed that $\lambda = 4c/3$.

(c) The multiplier λ is the rate of change of $m(c)$ as c increases and the constraint moves: that is, $\lambda = m'(c)$.

37. (a) The objective function is the complementary energy, $\dfrac{f_1^2}{2k_1} + \dfrac{f_2^2}{2k_2}$, and the constraint is $f_1 + f_2 = mg$. The Lagrangian function is

$$\mathcal{L}(f_1, f_2, \lambda) = \frac{f_1^2}{2k_1} + \frac{f_2^2}{2k_2} - \lambda(f_1 + f_2 - mg).$$

We look for solutions to the system of equations we get from grad $\mathcal{L} = \vec{0}$:

$$\frac{\partial \mathcal{L}}{\partial f_1} = \frac{f_1}{k_1} - \lambda = 0$$

$$\frac{\partial \mathcal{L}}{\partial f_2} = \frac{f_2}{k_2} - \lambda = 0$$

$$\frac{\partial \mathcal{L}}{\partial \lambda} = -(f_1 + f_2 - mg) = 0.$$

Combining $\dfrac{\partial \mathcal{L}}{\partial f_1} - \dfrac{\partial \mathcal{L}}{\partial f_2} = \dfrac{f_1}{k_1} - \dfrac{f_2}{k_2} = 0$ with $\dfrac{\partial \mathcal{L}}{\partial \lambda} = 0$ gives the two equation system

$$\frac{f_1}{k_1} - \frac{f_2}{k_2} = 0$$

$$f_1 + f_2 = mg.$$

Substituting $f_2 = mg - f_1$ into the first equation leads to

$$f_1 = \frac{k_1}{k_1 + k_2} mg$$

$$f_2 = \frac{k_2}{k_1 + k_2} mg.$$

(b) Hooke's Law states that for a spring

$$\text{Force of spring} = \text{Spring constant} \cdot \text{Distance stretched or compressed from equilibrium}.$$

Since $f_1 = k_1 \cdot \lambda$ and $f_2 = k_2 \cdot \lambda$, the Lagrange multiplier λ equals the distance the mass stretches the top spring and compresses the lower spring.

41. (a) The objective function $f(x, y) = px + qy$ gives the cost to buy x units of input 1 at unit price p and y units of input 2 at unit price q.

The constraint $g(x, y) = u$ tells us that we are only considering the cost of inputs x and y that can be used to produce quantity u of the product.

Thus the number $C(p, q, u)$ gives the minimum cost to the company of producing quantity u if the inputs it needs have unit prices p and q.

(b) The Lagrangian function is

$$\mathcal{L}(x, y, \lambda) = px + qy - \lambda(xy - u).$$

We look for solutions to the system of equations we get from grad $\mathcal{L} = \vec{0}$:

$$\frac{\partial \mathcal{L}}{\partial x} = p - \lambda y = 0$$

$$\frac{\partial \mathcal{L}}{\partial y} = q - \lambda x = 0$$

$$\frac{\partial \mathcal{L}}{\partial \lambda} = -(xy - u) = 0.$$

We see that $\lambda = p/y = q/x$ so $y = px/q$. Substituting for y in the constraint $xy = u$ leads to $x = \sqrt{qu/p}$, $y = \sqrt{pu/q}$ and $\lambda = \sqrt{pq/u}$. The minimum cost is thus

$$C(p, q, u) = p\sqrt{\frac{qu}{p}} + q\sqrt{\frac{pu}{q}} = 2\sqrt{pqu}.$$

45. (a) If the prices are p_1 and p_2 and the budget is b, the quantities consumed are constrained by

$$p_1 x_1 + p_2 x_2 = b.$$

We want to maximize

$$u(x_1, x_2) = a \ln x_1 + (1 - a) \ln x_2$$

subject to the constraint

$$p_1 x_1 + p_2 x_2 = b.$$

Using Lagrange multipliers, we solve

$$\frac{\partial u}{\partial x_1} = \frac{a}{x_1} = \lambda p_1$$

$$\frac{\partial u}{\partial x_2} = \frac{1 - a}{x_2} = \lambda p_2,$$

giving $x_1 = a/(\lambda p_1)$ and $x_2 = (1 - a)/(\lambda p_2)$. Substituting into the constraint, we get

$$\frac{a}{\lambda} + \frac{(1 - a)}{\lambda} = b$$

so

$$\lambda = \frac{1}{b}.$$

Thus

$$x_1 = \frac{ab}{p_1} \qquad x_2 = \frac{(1 - a)b}{p_2}$$

so the maximum satisfaction is given by

$$S = u(x_1, x_2) = u\left(\frac{ab}{p_1}, \frac{(1 - a)b}{p_2}\right) = a \ln\left(\frac{ab}{p_1}\right) + (1 - a) \ln\left(\frac{(1 - a)b}{p_2}\right)$$

$$= a \ln a + a \ln b - a \ln p_1 + (1 - a) \ln(1 - a) + (1 - a) \ln b - (1 - a) \ln p_2$$

$$= a \ln a + (1 - a) \ln(1 - a) + \ln b - a \ln p_1 - (1 - a) \ln p_2.$$

(b) We want to calculate the value of b needed to achieve $u(x_1, x_2) = c$. Thus, we solve for b in the equation

$$c = a \ln a + (1 - a) \ln(1 - a) + \ln b - a \ln p_1 - (1 - a) \ln p_2.$$

Since
$$\ln b = c - a \ln a - (1-a)\ln(1-a) + a \ln p_1 + (1-a)\ln p_2,$$
we have
$$b = \frac{e^c \cdot e^{a \ln p_1} \cdot e^{(1-a)\ln p_2}}{e^{a \ln a} \cdot e^{(1-a)\ln(1-a)}} = \frac{e^c p_1^a p_2^{1-a}}{a^a (1-a)^{(1-a)}}.$$

Solutions for Chapter 15 Review

Exercises

1. The critical points of f are obtained by solving $f_x = f_y = 0$, that is
$$f_x(x,y) = 2y^2 - 2x = 0 \quad \text{and} \quad f_y(x,y) = 4xy - 4y = 0,$$
so
$$2(y^2 - x) = 0 \quad \text{and} \quad 4y(x-1) = 0$$
The second equation gives either $y = 0$ or $x = 1$. If $y = 0$ then $x = 0$ by the first equation, so $(0,0)$ is a critical point. If $x = 1$ then $y^2 = 1$ from which $y = 1$ or $y = -1$, so two further critical points are $(1,-1)$, and $(1,1)$.

Since
$$D = f_{xx}f_{yy} - (f_{xy})^2 = (-2)(4x - 4) - (4y)^2 = 8 - 8x - 16y^2,$$
we have
$$D(0,0) = 8 > 0, \quad D(1,1) = D(1,-1) = -16 < 0,$$
and $f_{xx} = -2 < 0$. Thus, $(0,0)$ is a local maximum; $(1,1)$ and $(1,-1)$ are saddle points.

5. The partial derivatives are
$$f_x = \cos x + \cos(x+y).$$
$$f_y = \cos y + \cos(x+y).$$

Setting $f_x = 0$ and $f_y = 0$ gives
$$\cos x = \cos y$$
For $0 < x < \pi$ and $0 < y < \pi$, $\cos x = \cos y$ only if $x = y$. Then, setting $f_x = f_y = 0$:
$$\cos x + \cos 2x = 0,$$
$$\cos x + 2\cos^2 x - 1 = 0,$$
$$(2\cos x - 1)(\cos x + 1) = 0.$$

So $\cos x = 1/2$ or $\cos x = -1$, that is $x = \pi/3$ or $x = \pi$. For the given domain $0 < x < \pi$, $0 < y < \pi$, we only consider the solution when $x = \pi/3$ then $y = x = \pi/3$. Therefore, the critical point is $\left(\frac{\pi}{3}, \frac{\pi}{3}\right)$.

Since
$$f_{xx}(x,y) = -\sin x - \sin(x+y) \quad f_{xx}\left(\tfrac{\pi}{3}, \tfrac{\pi}{3}\right) = -\sin\tfrac{\pi}{3} - \sin\tfrac{2\pi}{3} = -\sqrt{3}$$
$$f_{xy}(x,y) = -\sin(x+y) \qquad\qquad f_{xy}\left(\tfrac{\pi}{3}, \tfrac{\pi}{3}\right) = -\sin\tfrac{2\pi}{3} \qquad = -\tfrac{\sqrt{3}}{2}$$
$$f_{yy}(x,y) = -\sin y - \sin(x+y) \quad f_{yy}\left(\tfrac{\pi}{3}, \tfrac{\pi}{3}\right) = -\sin\tfrac{\pi}{3} - \sin\tfrac{2\pi}{3} = -\sqrt{3}$$
the discriminant is
$$D(x,y) = f_{xx}f_{yy} - f_{xy}^2$$
$$= (-\sqrt{3})(-\sqrt{3}) - (-\tfrac{\sqrt{3}}{2})^2 = \tfrac{9}{4} > 0.$$

Since $f_{xx}\left(\tfrac{\pi}{3}, \tfrac{\pi}{3}\right) = -\sqrt{3} < 0$, $\left(\tfrac{\pi}{3}, \tfrac{\pi}{3}\right)$ is a local maximum.

9. The partial derivatives are
$$f_x = y + \frac{1}{x}, f_y = x + 2y.$$

For critical points, solve $f_x = 0$ and $f_y = 0$ simultaneously. From $f_y = x + 2y = 0$ we get that $x = -2y$. Substituting into $f_x = 0$, we have
$$y + \frac{1}{x} = y - \frac{1}{2y} = \frac{1}{y}\left(y^2 - \frac{1}{2}\right) = 0$$

Since $\frac{1}{y} \neq 0$, $y^2 - \frac{1}{2} = 0$, therefore
$$y = \pm\frac{1}{\sqrt{2}} = \pm\frac{\sqrt{2}}{2},$$

and $x = \mp\sqrt{2}$. So the critical points are $\left(-\sqrt{2}, \frac{\sqrt{2}}{2}\right)$ and $\left(\sqrt{2}, -\frac{\sqrt{2}}{2}\right)$. But x must be greater than 0, so $\left(-\sqrt{2}, \frac{\sqrt{2}}{2}\right)$ is not in the domain.

The contour diagram for f in Figure 15.6 (drawn by computer), shows that $\left(\sqrt{2}, -\frac{\sqrt{2}}{2}\right)$ is a saddle point of $f(x, y)$.

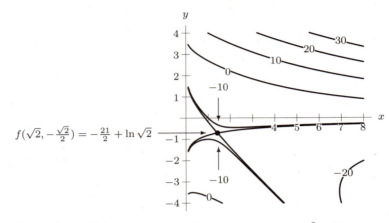

$$f\left(\sqrt{2}, -\frac{\sqrt{2}}{2}\right) = -\frac{21}{2} + \ln\sqrt{2}$$

Figure 15.6: Contour map of $f(x, y) = xy + \ln x + y^2 - 10$

We can also see that $\left(\sqrt{2}, -\frac{\sqrt{2}}{2}\right)$ is a saddle point analytically.

Since $f_{xx} = -\frac{1}{x^2}, f_{yy} = 2, f_{xy} = 1$, the discriminant is:

$$D(x, y) = f_{xx}f_{yy} - f_{xy}^2$$
$$= -\frac{2}{x^2} - 1.$$

$D\left(\sqrt{2}, -\frac{\sqrt{2}}{2}\right) = -2 < 0$, so $\left(\sqrt{2}, -\frac{\sqrt{2}}{2}\right)$ is a saddle point.

13. The objective function is $f(x, y) = x^2 - xy + y^2$ and the equation of constraint is $g(x, y) = x^2 - y^2 = 1$. The gradients of f and g are

$$\nabla f(x, y) = (2x - y)\vec{i} + (-x + 2y)\vec{j},$$
$$\nabla g(x, y) = 2x\vec{i} - 2y\vec{j}.$$

Therefore the equation $\nabla f(x, y) = \lambda \nabla g(x, y)$ gives

$$2x - y = 2\lambda x$$
$$-x + 2y = -2\lambda y$$
$$x^2 - y^2 = 1.$$

Let us suppose that $\lambda = 0$. Then $2x = y$ and $2y = x$ give $x = y = 0$. But $(0, 0)$ is not a solution of the third equation, so we conclude that $\lambda \neq 0$. Now let's multiply the first two equations

$$-2\lambda y(2x - y) = 2\lambda x(-x + 2y).$$

As $\lambda \neq 0$, we can cancel it in the equation above and after doing the algebra we get

$$x^2 - 4xy + y^2 = 0$$

which gives $x = (2 + \sqrt{3})y$ or $x = (2 - \sqrt{3})y$.

If $x = (2 + \sqrt{3})y$, the third equation gives

$$(2 + \sqrt{3})^2 y^2 - y^2 = 1$$

so $y \approx \pm 0.278$ and $x \approx \pm 1.038$. These give the critical points $(1.038, 0.278)$, $(-1.038, -0.278)$.

If $x = (2 - \sqrt{3})y$, from the third equation we get

$$(2 - \sqrt{3})^2 y^2 - y^2 = 1.$$

But $(2 - \sqrt{3})^2 - 1 \approx -0.928 < 0$ so the equation has no solution. Evaluating f gives

$$f(1.038, 0.278) = f(-1.038, -0.278) \approx 0.866$$

Since $y \to \infty$ on the constraint, rewriting f as

$$f(x, y) = \left(x - \frac{y}{2}\right)^2 + \frac{3}{4}y^2$$

shows that f has no maximum on the constraint. The minimum value of f is 0.866. See Figure 15.7.

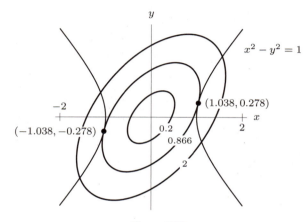

Figure 15.7

17. Our objective function is $f(x, y, z) = x^2 - 2y + 2z^2$ and our equation of constraint is $g(x, y, z) = x^2 + y^2 + z^2 - 1 = 0$. To optimize $f(x, y, z)$ with Lagrange multipliers, we solve $\nabla f(x, y, z) = \lambda \nabla g(x, y, z)$ subject to $g(x, y, z) = 0$. The gradients of f and g are

$$\nabla f(x, y, z) = 2x\vec{i} - 2\vec{j} + 4z\vec{k},$$
$$\nabla g(x, y) = 2x\vec{i} + 2y\vec{j} + 2z\vec{k}.$$

We get,

$$x = \lambda x$$
$$-1 = \lambda y$$
$$2z = \lambda z$$
$$x^2 + y^2 + z^2 = 1.$$

From the first equation we get $x = 0$ or $\lambda = 1$.

If $x = 0$ we have

$$-1 = \lambda y$$
$$2z = \lambda z$$
$$y^2 + z^2 = 1.$$

From the second equation $z = 0$ or $\lambda = 2$. So if $z = 0$, we have $y = \pm 1$ and we get the solutions $(0, 1, 0), (0, -1, 0)$. If $z \neq 0$ then $\lambda = 2$ and $y = -\frac{1}{2}$. So $z^2 = \frac{3}{4}$ which gives the solutions $(0, -\frac{1}{2}, \frac{\sqrt{3}}{2}), (0, -\frac{1}{2}, -\frac{\sqrt{3}}{2})$.

If $x \neq 0$, then $\lambda = 1$, so $y = -1$, which implies, from the equation $x^2 + y^2 + z^2 = 1$, that $x = 0$, which contradicts the assumption.

Since the constraint is closed and bounded, maximum and minimum values of f subject to the constraint exist. Therefore, evaluating f at the critical points, we get $f(0, 1, 0) = -2$, $f(0, -1, 0) = 2$ and $f(0, -\frac{1}{2}, \frac{\sqrt{3}}{2}) = f(0, -\frac{1}{2}, -\frac{\sqrt{3}}{2}) = 4$. So the maximum value of f is 4 and the minimum is -2.

21. The region $x + y \geq 1$ is the shaded half plane (including the line $x + y = 1$) shown in Figure 15.8.

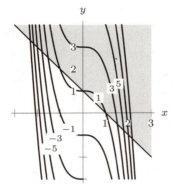

y

Figure 15.8

Let's look for the critical points of f in the interior of the region. As

$$f_x = 3x^2$$
$$f_y = 1$$

there are no critical points inside the shaded region. Now let's find the extrema of f on the boundary of our region. We want the extrema of $f(x,y) = x^3 + y$ subject to the constraint $g(x,y) = x + y - 1 = 0$. We use Lagrange multipliers

$$\text{grad } f = \lambda \text{ grad } g \quad \text{and} \quad x + y = 1,$$

which give

$$3x^2 = \lambda$$
$$1 = \lambda$$
$$x + y = 1.$$

From the first two equations we get $3x^2 = 1$, so the solutions are

$$(\frac{1}{\sqrt{3}}, 1 - \frac{1}{\sqrt{3}}) \quad \text{and} \quad (-\frac{1}{\sqrt{3}}, 1 + \frac{1}{\sqrt{3}}).$$

Evaluating f at these points we get

$$f(\frac{1}{\sqrt{3}}, 1 - \frac{1}{\sqrt{3}}) = 1 - \frac{2}{3\sqrt{3}}$$
$$f(-\frac{1}{\sqrt{3}}, 1 + \frac{1}{\sqrt{3}}) = 1 + \frac{2}{3\sqrt{3}}.$$

From the contour diagram in Figure 15.8, we see that $(\frac{1}{\sqrt{3}}, 1 - \frac{1}{\sqrt{3}})$ is a local minimum and $(-\frac{1}{\sqrt{3}}, 1 + \frac{1}{\sqrt{3}})$ is a local maximum of f on $x + y = 1$. Are they global extrema as well?

If we take x very big and $y = 1 - x$ then $f(x,y) = x^3 + y = x^3 - x + 1$ which can be made as big as we want (if we choose x big enough). So there will be no global maximum.

Similarly, taking x negative with big absolute value and $y = 1 - x$, $f(x,y) = x^3 + y = x^3 - x + 1$ can be made as small as we want (if we choose x small enough). So there is no global minimum. This can also be seen from Figure 15.8.

25. If $xy = 10$, then $f(x,y) = x^2 - y^2 = x^2 - 100/x^2$ and x can take any nonzero value. Since

$$\lim_{x \to \infty} \left(x^2 - \frac{100}{x^2} \right) = \infty,$$

we see f has no maximum on the constraint. Since

$$\lim_{x \to 0} \left(x^2 - \frac{100}{x^2} \right) = -\infty,$$

we see f has no minimum on the constraint.

Problems

29. (a) (i) Suppose $N = kA^p$. Then the rule of thumb tells us that if A is multiplied by 10, the value of N doubles. Thus

$$2N = k(10A)^p = k10^p A^p.$$

Thus, dividing by $N = kA^p$, we have

$$2 = 10^p$$

so taking logs to base 10 we have

$$p = \log 2 = 0.3010.$$

(where $\log 2$ means $\log_{10} 2$). Thus,

$$N = kA^{0.3010}.$$

(ii) Taking natural logs gives

$$\ln N = \ln(kA^p)$$
$$\ln N = \ln k + p \ln A$$
$$\ln N \approx \ln k + 0.301 \ln A$$

Thus, $\ln N$ is a linear function of $\ln A$.

(b) Table 15.1 contains the natural logarithms of the data:

Table 15.1 $\ln N$ and $\ln A$

Island	$\ln A$	$\ln N$
Redonda	1.1	1.6
Saba	3.0	2.2
Montserrat	2.3	2.7
Puerto Rico	9.1	4.3
Jamaica	9.3	4.2
Hispaniola	11.2	4.8
Cuba	11.6	4.8

Using a least squares fit we find the line:

$$\ln N = 1.20 + 0.32 \ln A$$

This yields the power function:

$$N = e^{1.20} A^{0.32} = 3.32 A^{0.32}$$

Since 0.32 is pretty close to $\log 2 \approx 0.301$, the answer does agree with the biological rule.

33. (a) The problem is to maximize

$$V = 1000 D^{0.6} N^{0.3}$$

subject to the budget constraint in dollars

$$40000D + 10000N \leq 600000$$

or (in thousand dollars)

$$40D + 10N \leq 600$$

(b) Let $B = 40D + 10N = 600$ (thousand dollars) be the budget constraint. At the optimum

$$\nabla V = \lambda \nabla B,$$

so $\quad \dfrac{\partial V}{\partial D} = \lambda \dfrac{\partial B}{\partial D} = 40\lambda$

$$\dfrac{\partial V}{\partial N} = \lambda \dfrac{\partial B}{\partial N} = 10\lambda.$$

Thus

$$\dfrac{\partial V/\partial D}{\partial V/\partial N} = 4.$$

Therefore, at the optimum point, the rate of increase in the number of visits with respect to an increase in the number of doctors is four times the corresponding rate for nurses. This factor of four is the same as the ratio of the salaries.

(c) Differentiating and setting $\nabla V = \lambda \nabla B$ yields

$$600D^{-0.4}N^{0.3} = 40\lambda$$
$$300D^{0.6}N^{-0.7} = 10\lambda$$

Thus, we get

$$\frac{600D^{-0.4}N^{0.3}}{40} = \lambda = \frac{300D^{0.6}N^{-0.7}}{10}$$

So

$$N = 2D.$$

To solve for D and N, substitute in the budget constraint:

$$600 - 40D - 10N = 0$$

$$600 - 40D - 10 \cdot (2D) = 0$$

So $D = 10$ and $N = 20$.

$$\lambda = \frac{600(10^{-0.4})(20^{0.3})}{40} \approx 14.67$$

Thus the clinic should hire 10 doctors and 20 nurses. With that staff, the clinic can provide

$$V = 1000(10^{0.6})(20^{0.3}) \approx 9,779 \text{ visits per year.}$$

(d) From part c), the Lagrange multiplier is $\lambda = 14.67$. At the optimum, the Lagrange multiplier tells us that about 14.67 extra visits can be generated through an increase of \$1,000 in the budget. (If we had written out the constraint in dollars instead of thousands of dollars, the Lagrange multiplier would tell us the number of extra visits per dollar.)

(e) The marginal cost, MC, is the cost of an additional visit. Thus, at the optimum point, we need the reciprocal of the Lagrange multiplier:

$$\text{MC} = \frac{1}{\lambda} \approx \frac{1}{14.67} \approx 0.068 \text{ (thousand dollars),}$$

that is, at the optimum point, an extra visit costs the clinic 0.068 thousand dollars, or \$68.

This production function exhibits declining returns to scale (e.g. doubling both inputs less than doubles output, because the two exponents add up to less than one). This means that for large V, increasing V will require increasing D and N by more than when V is small. Thus the cost of an additional visit is greater for large V than for small. In other words, the marginal cost will rise with the number of visits.

37. The objective function is

$$f(x, y, z) = \sqrt{(x - a)^2 + (y - b)^2 + (z - c)^2},$$

and the constraint is

$$g(x, y, z) = Ax + By + Cz + D = 0.$$

Partial derivatives of f and g are

$$f_x = \frac{\frac{1}{2} \cdot 2 \cdot (x - a)}{f(x, y, z)} = \frac{x - a}{f(x, y, z)},$$

$$f_y = \frac{\frac{1}{2} \cdot 2 \cdot (y - b)}{f(x, y, z)} = \frac{y - b}{f(x, y, z)},$$

$$f_z = \frac{\frac{1}{2} \cdot 2 \cdot (z - c)}{f(x, y, z)} = \frac{z - c}{f(x, y, z)},$$

$$g_x = A, \quad g_y = B, \quad \text{and} \quad g_z = C.$$

Using Lagrange multipliers, we need to solve the equations

$$\text{grad } f = \lambda \, \text{grad } g$$

where grad $f = f_x \vec{i} + f_y \vec{j} + f_z \vec{k}$ and grad $g = g_x \vec{i} + g_y \vec{j} + g_z \vec{k}$. This gives a system of equations:

$$\frac{x - a}{f(x, y, z)} = \lambda A$$

$$\frac{y - b}{f(x, y, z)} = \lambda B$$

$$\frac{z - c}{f(x, y, z)} = \lambda C$$

$$Ax + By + Cz + D = 0.$$

Now $\frac{x-a}{A} = \frac{y-b}{B} = \frac{z-c}{C} = \lambda f(x, y, z)$ gives

$$x = \frac{A}{B}(y - b) + a,$$

$$z = \frac{C}{B}(y - b) + c,$$

Substitute into the constraint,

$$A\left(\frac{A}{B}(y - b) + a\right) + By + C\left(\frac{C}{B}(y - b) + c\right) + D = 0,$$

$$\left(\frac{A^2}{B} + B + \frac{C^2}{B}\right)y = \frac{A^2}{B}b - Aa + \frac{C^2}{B}b - Cc - D.$$

Hence

$$y = \frac{(A^2 + C^2)b - B(Aa + Cc + D)}{A^2 + B^2 + C^2},$$

$$y - b = \frac{-B(Aa + Bb + Cc + D)}{A^2 + B^2 + C^2}$$

$$x - a = \frac{A}{B}(y - b)$$

$$= \frac{-A(Aa + Bb + Cc + D)}{A^2 + B^2 + C^2}$$

$$z - c = \frac{C}{B}(y - b)$$

$$= \frac{-C(Aa + Bb + Cc + D)}{A^2 + B^2 + C^2}$$

Thus the minimum $f(x, y, z)$ is

$$f(x, y, z) = \sqrt{(x - a)^2 + (y - b)^2 + (z - c)^2}$$

$$= \left[\left(\frac{-A(Aa + Bb + Cc + D)}{A^2 + B^2 + C^2}\right)^2 + \left(\frac{-B(Aa + Bb + Cc + D)}{A^2 + B^2 + C^2}\right)^2\right.$$

$$\left. + \left(\frac{-C(Aa + Bb + Cc + D)}{A^2 + B^2 + C^2}\right)^2\right]^{1/2}$$

$$= \frac{|Aa + Bb + Cc + D|}{\sqrt{A^2 + B^2 + C^2}}.$$

The geometric meaning is finding the shortest distance from a point (a, b, c) to the plane $Ax + By + Cz + D = 0$.

41. Cost of production, C, is given by $C = p_1 W + p_2 K = b$. At the optimal point, $\nabla q = \lambda \nabla C$.
Since $\nabla q = \left(c(1 - a)W^{-a}K^a\right)\vec{i} + \left(caW^{1-a}K^{a-1}\right)\vec{j}$ and $\nabla C = p_1\vec{i} + p_2\vec{j}$, we get

$$c(1 - a)W^{-a}K^a = \lambda p_1 \quad \text{and} \quad caW^{1-a}K^{a-1} = \lambda p_2.$$

Now, marginal productivity of labor is given by $\frac{\partial q}{\partial W} = c(1 - a)W^{-a}K^a$ and marginal productivity of capital is given by $\frac{\partial q}{\partial K} = caW^{1-a}K^{a-1}$, so their ratio is given by

$$\frac{\frac{\partial q}{\partial W}}{\frac{\partial q}{\partial K}} = \frac{c(1 - a)W^{-a}K^a}{caW^{1-a}K^{a-1}} = \frac{\lambda p_1}{\lambda p_2} = \frac{p_1}{p_2}$$

which is the ratio of the cost of one unit of labor to the cost of one unit of capital.

45. You should try to anticipate your opponent's choice. After you choose a value λ, your opponent will use calculus to find the point (x, y) that maximizes the function $f(x, y) = 10 - x^2 - y^2 - 2x - \lambda(2x + 2y)$. At that point, we have $f_x = -2x - 2 - 2\lambda = 0$ and $f_y = -2y - 2\lambda = 0$, so your opponent will choose $x = -1 - \lambda$ and $y = -\lambda$. This gives a value $\mathcal{L}(-1 - \lambda, -\lambda, \lambda) = 10 - (-1 - \lambda)^2 - (-\lambda)^2 - 2(-1 - \lambda) - \lambda(2(-1 - \lambda) + 2(-\lambda)) = 11 + 2\lambda + 2\lambda^2$ which you want to make as small as possible. You should choose λ to minimize the function $h(\lambda) = 11 + 2\lambda + 2\lambda^2$. You choose λ so that $h'(\lambda) = 2 + 4\lambda = 0$, or $\lambda = -1/2$. Your opponent then chooses $(x, y) = (-1 - \lambda, -\lambda) = (-1/2, 1/2)$, giving a final score of $\mathcal{L}(-1/2, 1/2, -1/2) = 10.5$. No choice of λ that you can make can force the value of $\mathcal{L}$ below 10.5. But your choice of $\lambda = -1/2$ makes it impossible for your opponent to force the value of $\mathcal{L}$ above 10.5.

CAS Challenge Problems

49. (a) We have grad $f = 3\vec{i} + 2\vec{j}$ and grad $g = (4x - 4y)\vec{i} + (-4x + 10y)\vec{j}$, so the Lagrange multiplier equations are

$$3 = \lambda(4x - 4y)$$
$$2 = \lambda(-4x + 10y)$$
$$2x^2 - 4xy + 5y^2 = 20$$

Solving these with a CAS we get $\lambda = -0.4005, x = -3.9532, y = -2.0806$ and $\lambda = 0.4005, x = 3.9532, y = 2.0806$. We have $f(-3.9532, -2.0806) = -11.0208$, and $f(3, 9532, 2.0806) = 21.0208$. The constraint equation is $2x^2 - 4xy + 5y^2 = 20$, or, completing the square, $2(x - y)^2 + 3y^2 = 20$. This has the shape of a skewed ellipse, so the constraint curve is bounded, and therefore the local maximum is a global maximum. Thus the maximum value is 21.0208.

(b) The maximum value on $g = 20.5$ is $\approx 21.0208 + 0.5(0.4005) = 21.2211$. The maximum value on $g = 20.2$ is $\approx 21.0208 + 0.2(0.4005) = 21.1008$.

(c) We use the same commands in the CAS from part (a), with 20 replaced by 20.5 and 20.2, and get the maximum values 21.2198 for $g = 20.5$ and 21.1007 for $g = 20.2$. These agree with the approximations we found in part (b) to 2 decimal places.

CHECK YOUR UNDERSTANDING

1. True. By definition, a critical point is either where the gradient of f is zero or does not exist.

5. True. The graph of this function is a cone that opens upward with its vertex at the origin.

9. False. For example, the linear function $f(x, y) = x + y$ has no local extrema at all.

13. False. For example, the linear function $f(x, y) = x + y$ has neither a global minimum or global maximum on all of 2-space.

17. False. On the given region the function f is always less than one. By picking points closer and closer to the circle $x^2 + y^2 = 1$ we can make f larger and larger (although never larger than one). There is no point in the open disk that gives f its largest value.

21. True. The point (a, b) must lie on the constraint $g(x, y) = c$, so $g(a, b) = c$.

25. False. Since grad f and grad g point in opposite directions, they are parallel. Therefore (a, b) could be a local maximum or local minimum of f constrained to $g = c$. However the information given is not enough to determine that it is a minimum. If the contours of g near (a, b) increase in the opposite direction as the contours of f, then at a point with grad $f(a, b) = \lambda$grad $g(a, b)$ we have $\lambda \leq 0$, but this can be a local maximum or minimum.

For example, $f(x, y) = 4 - x^2 - y^2$ has a local maximum at $(1, 1)$ on the constraint $g(x, y) = x + y = 2$. Yet at this point, grad $f = -2\vec{i} - 2\vec{j}$ and grad $g = \vec{i} + \vec{j}$, so grad f and grad g point in opposite directions.

29. True. Since $f(a, b) = M$, we must satisfy the Lagrange conditions that $f_x(a, b) = \lambda g(a, b)$ and $f_y(a, b) = \lambda g_y(a, b)$, for some λ. Thus $f_x(a, b)/f_y(a, b) = g_x(a, b)/g_y(a, b)$.

33. False. The value of λ at a minimum point gives the proportional change in m for a change in c. If $\lambda > 0$ and the change in c is positive, the change in m will also be positive.

CHAPTER SIXTEEN

Solutions for Section 16.1

Exercises

1. Mark the values of the function on the plane, as shown in Figure 16.1, so that you can guess respectively at the smallest and largest values the function takes on each small rectangle.

$$\text{Lower sum} = \sum f(x_i, y_i) \Delta x \Delta y$$
$$= 4\Delta x \Delta y + 6\Delta x \Delta y + 3\Delta x \Delta y + 4\Delta x \Delta y$$
$$= 17\Delta x \Delta y$$
$$= 17(0.1)(0.2) = 0.34.$$

$$\text{Upper sum} = \sum f(x_i, y_i) \Delta x \Delta y$$
$$= 7\Delta x \Delta y + 10\Delta x \Delta y + 6\Delta x \Delta y + 8\Delta x \Delta y$$
$$= 31\Delta x \Delta y$$
$$= 31(0.1)(0.2) = 0.62.$$

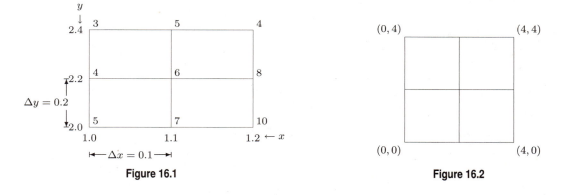

Figure 16.1   **Figure 16.2**

5. (a) If we take the partition of R consisting of just R itself, we get

$$\text{Lower bound for integral} = \min_R f \cdot A_R = 0 \cdot (4 - 0)(4 - 0) = 0.$$

Similarly, we get

$$\text{Upper bound for integral} = \max_R f \cdot A_R = 4 \cdot (4 - 0)(4 - 0) = 64.$$

(b) The estimates asked for are just the upper and lower sums. We partition R into subrectangles $R_{(a,b)}$ of width 2 and height 2, where (a, b) is the lower-left corner of $R_{(a,b)}$. The subrectangles are then $R_{(0,0)}$, $R_{(2,0)}$, $R_{(0,2)}$, and $R_{(2,2)}$, as in Figure 16.2. Then we find the lower sum

$$\text{Lower sum} = \sum_{(a,b)} A_{R_{(a,b)}} \cdot \min_{R_{(a,b)}} f = \sum_{(a,b)} 4 \cdot (\text{Min of } f \text{ on } R_{(a,b)})$$
$$= 4 \sum_{(a,b)} (\text{Min of } f \text{ on } R_{(a,b)})$$
$$= 4(f(0,0) + f(2,0) + f(0,2) + f(2,2))$$
$$= 4(\sqrt{0 \cdot 0} + \sqrt{2 \cdot 0} + \sqrt{0 \cdot 2} + \sqrt{2 \cdot 2})$$
$$= 8.$$

Similarly, the upper sum is

$$\text{Upper sum} = 4 \sum_{(a,b)} (\text{Max of } f \text{ on } R_{(a,b)})$$
$$= 4(f(2,2) + f(4,2) + f(2,4) + f(4,4))$$
$$= 4(\sqrt{2 \cdot 2} + \sqrt{4 \cdot 2} + \sqrt{2 \cdot 4} + \sqrt{4 \cdot 4})$$
$$= 24 + 16\sqrt{2} \approx 46.63.$$

The upper sum is an overestimate and the lower sum is an underestimate, so we can get a better estimate by averaging them to get $16 + 8\sqrt{2} \approx 27.3$.

Problems

9. The function being integrated is $f(x, y) = 1$, which is positive everywhere. Thus, its integral over any region is positive.

13. The region D is symmetric both with respect to x and y axes. The function being integrated is $f(x, y) = 5x$, which is an odd function in x. Since D is symmetric with respect to x, the contributions to the integral cancel out. Thus, the integral of the function over the region D is zero.

17. The function being integrated, $f(x, y) = y - y^3$ is always negative in the region B since in that region $-1 < y < 0$ and $|y^3| < |y|$. Thus, the integral is negative.

21. The function $f(x, y) = e^x$ is positive for any value of x. Thus, its integral is always positive for any region, such as D, with nonzero area.

25. The question is asking which graph has more volume under it, and from inspection, it appears that it would be the graph for the mosquitos.

29. Let R be the region $0 \le x \le 60, \quad 0 \le y \le 8$. Then

$$\text{Volume} = \int_R w(x, y) \, dA$$

Lower estimate: $10 \cdot 2(1+4+8+10+10+8+0+3+4+6+6+4+0+1+2+3+3+2+0+0+1+1+1+1) = 1580$.
Upper estimate:
$10 \cdot 2(8+13+16+17+17+16+4+8+10+11+11+10+3+4+6+7+7+6+1+2+3+4+4+3) = 3820$.
The average of the two estimates is 2700 cubic feet.

Solutions for Section 16.2

Exercises

1. See Figure 16.3.

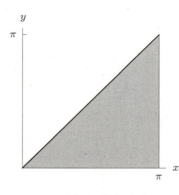

Figure 16.3

5. We evaluate the inside integral first:

$$\int_0^4 (4x + 3y)\,dx = (2x^2 + 3yx)\Big|_0^4 = 32 + 12y.$$

Therefore, we have

$$\int_0^3 \int_0^4 (4x + 3y)\,dxdy = \int_0^3 (32 + 12y)\,dy = (32y + 6y^2)\Big|_0^3 = 150.$$

9. Calculating the inner integral first, we have

$$\int_0^1 \int_0^1 ye^{xy}\,dx\,dy = \int_0^1 \left(e^{xy}\Big|_0^1\right) dy = \int_0^1 \left(e^y - e^0\right) dy = \int_0^1 (e^y - 1)\,dy = (e^y - y)\Big|_0^1 = e^1 - 1 - (e^0 - 0) = e - 2.$$

13. This region lies between $x = 0$ and $x = 4$ and between the lines $y = 3x$ and $y = 12$, and so the iterated integral is

$$\int_0^4 \int_{3x}^{12} f(x, y)\,dydx.$$

Alternatively, we could have set up the integral as follows:

$$\int_0^{12} \int_0^{y/3} f(x, y)\,dxdy.$$

17. The line connecting $(1, 0)$ and $(4, 1)$ is

$$y = \frac{1}{3}(x - 1)$$

So the integral is

$$\int_1^4 \int_{(x-1)/3}^2 f\,dy\,dx$$

21.

$$\int_1^4 \int_{\sqrt{y}}^y x^2 y^3\,dxdy = \int_1^4 y^3 \frac{x^3}{3}\Big|_{\sqrt{y}}^y dy$$

$$= \frac{1}{3} \int_1^4 (y^6 - y^{\frac{9}{2}})\,dy$$

$$= \frac{1}{3}\left(\frac{y^7}{7} - \frac{y^{11/2}}{11/2}\right)\Big|_1^4$$

$$= \frac{1}{3}\left[\left(\frac{4^7}{7} - \frac{4^{11/2} \times 2}{11}\right) - \left(\frac{1}{7} - \frac{2}{11}\right)\right] \approx 656.082$$

See Figure 16.4.

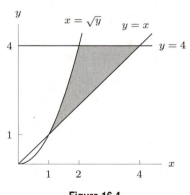

Figure 16.4

25. In the other order, the integral is

$$\int_0^1 \int_0^2 \sqrt{x+y}\, dy\, dx.$$

First we keep x fixed and calculate the inside integral with respect to y:

$$\int_0^2 \sqrt{x+y}\, dy = \frac{2}{3}(x+y)^{3/2}\Big|_{y=0}^{y=2}$$
$$= \frac{2}{3}\left[(x+2)^{3/2} - x^{3/2}\right].$$

Then the outside integral becomes

$$\int_0^1 \frac{2}{3}\left[(x+2)^{3/2} - x^{3/2}\right]\, dx = \frac{2}{3}\left[\frac{2}{5}(x+2)^{5/2} - \frac{2}{5}x^{5/2}\right]\Big|_0^1$$
$$= \frac{2}{3}\cdot\frac{2}{5}\left[3^{5/2} - 1 - 2^{5/2}\right] = 2.38176$$

Note that the answer is the same as the one we got in Exercise 24.

Problems

29. The region is bounded by $x = 1$, $x = 4$, $y = 2$, and $y = 2x$. Thus

$$\text{Volume} = \int_1^4 \int_2^{2x} (6x^2 y)\, dy\, dx.$$

To evaluate this integral, we evaluate the inside integral first:

$$\int_2^{2x} (6x^2 y)\, dy = (3x^2 y^2)\Big|_2^{2x} = 3x^2(2x)^2 - 3x^2(2^2) = 12x^4 - 12x^2.$$

Therefore, we have

$$\int_1^4 \int_2^{2x} (6x^2 y)\, dy\, dx = \int_1^4 (12x^4 - 12x^2)\, dx = \left(\frac{12}{5}x^5 - 4x^3\right)\Big|_1^4 = 2203.2.$$

The volume of this object is 2203.2.

33. (a) The line $x = y/2$ is the line $y = 2x$, and $y = x$ and $y = 2x$ intersect at $x = 0$. Thus, R is the shaded region in Figure 16.5. One expression for the integral is

$$\int_R f\, dA = \int_0^3 \int_x^{2x} x^2 e^{x^2}\, dy\, dx.$$

Another expression is obtained by reversing the order of integration. When we do this, it is necessary to split R into two regions on a line parallel to the x-axis along the point of intersection of $y = x$ and $x = 3$; this line is $y = 3$. Then we obtain

$$\int_R f\, dA = \int_0^3 \int_{y/2}^y x^2 e^{x^2}\, dx\, dy + \int_3^6 \int_{y/2}^3 x^2 e^{x^2}\, dx\, dy.$$

(b) We evaluate the first integral. Integrating with respect to y first:

$$\int_0^3 \int_x^{2x} x^2 e^{x^2}\, dy\, dx = \int_0^3 (x^2 e^{x^2} y)\Big|_x^{2x}\, dx = \int_0^3 x^3 e^{x^2}\, dx.$$

We use integration by parts with $u = x^2$, $v' = xe^{x^2}$. Then $u' = 2x$ and $v = \frac{1}{2}e^{x^2}$, so

$$\int_0^3 x^3 e^{x^2}\, dx = \frac{1}{2}x^2 e^{x^2}\Big|_0^3 - \int_0^3 xe^{x^2}\, dx = \frac{1}{2}x^2 e^{x^2}\Big|_0^3 - \frac{1}{2}e^{x^2}\Big|_0^3$$
$$= \left(\frac{1}{2}(9)e^9 - 0\right) - \left(\frac{1}{2}e^9 - \frac{1}{2}\right) = \frac{1}{2} + 4e^9.$$

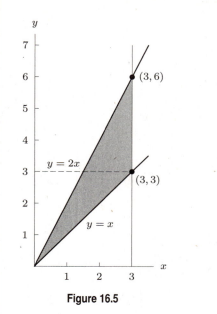

Figure 16.5

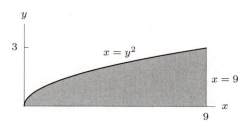

Figure 16.6

37. As given, the region of integration is as shown in Figure 16.6.

Reversing the limits gives

$$\int_0^9 \int_0^{\sqrt{x}} y \sin (x^2)\, dy dx = \int_0^9 \left(\frac{y^2 \sin (x^2)}{2} \Big|_0^{\sqrt{x}} \right) dx$$

$$= \frac{1}{2} \int_0^9 x \sin (x^2)\, dx$$

$$= -\frac{\cos (x^2)}{4} \Big|_0^9$$

$$= \frac{1}{4} - \frac{\cos (81)}{4} = 0.056.$$

41. (a) The contour $f(x, y) = 1$ lies in the xy-plane and has equation

$$2e^{-(x-1)^2 - y^2} = 1,$$

so

$$-(x-1)^2 - y^2 = \ln(1/2)$$
$$(x-1)^2 + y^2 = \ln 2 = 0.69.$$

This is the equation of a circle centered at $(1, 0)$ in the xy-plane.

Other contours are of the form

$$2e^{-(x-1)^2 - y^2} = c$$
$$-(x-1)^2 - y^2 = \ln(c/2).$$

Thus, all the contours are circles centered at the point $(1, 0)$.

(b) The cross-section has equation $z = f(1, y) = e^{-y^2}$. If $x = 1$, the base region in the xy-plane extends from $y = -\sqrt{3}$ to $y = \sqrt{3}$. See Figure 16.7, which shows the circular region below W in the xy-plane. So

$$\text{Area} = \int_{-\sqrt{3}}^{\sqrt{3}} e^{-y^2}\, dy.$$

(c) Slicing parallel to the y-axis, we get

$$\text{Volume} = \int_{-2}^{2} \int_{-\sqrt{4-x^2}}^{\sqrt{4-x^2}} e^{-(x-1)^2 - y^2} \, dy \, dx.$$

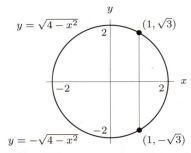

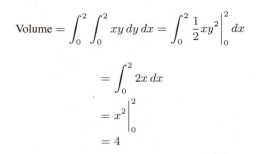

Figure 16.7: Region beneath W in the
xy-plane

45.

$$\text{Volume} = \int_{0}^{2} \int_{0}^{2} xy \, dy \, dx = \int_{0}^{2} \frac{1}{2} xy^2 \Big|_{0}^{2} \, dx$$

$$= \int_{0}^{2} 2x \, dx$$

$$= x^2 \Big|_{0}^{2}$$

$$= 4$$

49. Let R be the triangle with vertices $(1,0)$, $(2,2)$ and $(0,1)$. Note that $(3x + 2y + 1) - (x + y) = 2x + y + 1 > 0$ for $x, y > 0$, so $z = 3x + 2y + 1$ is above $z = x + y$ on R. We want to find

$$\text{Volume} = \int_{R} ((3x + 2y + 1) - (x + y)) \, dA = \int_{R} (2x + y + 1) \, dA.$$

We need to express this in terms of double integrals.

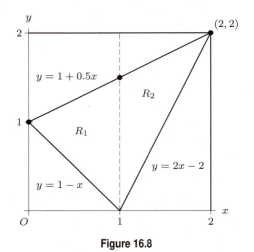

Figure 16.8

To do this, divide R into two regions with the line $x = 1$ to make regions R_1 for $x \leq 1$ and R_2 for $x \geq 1$. See Figure 16.8. We want to find

$$\int_R (2x + y + 1) \, dA = \int_{R_1} (2x + y + 1) \, dA + \int_{R_2} (2x + y + 1) \, dA.$$

Note that the line connecting $(0, 1)$ and $(1, 0)$ is $y = 1 - x$, and the line connecting $(0, 1)$ and $(2, 2)$ is $y = 1 + 0.5x$. So

$$\int_{R_1} (2x + y + 1) \, dA = \int_0^1 \int_{1-x}^{1+0.5x} (2x + y + 1) \, dy \, dx.$$

The line between $(1, 0)$ and $(2, 2)$ is $y = 2x - 2$, so

$$\int_{R_2} (2x + y + 1) \, dA = \int_1^2 \int_{2x-2}^{1+0.5x} (2x + y + 1) \, dy \, dx.$$

We can now compute the double integral for R_1:

$$\int_0^1 \int_{1-x}^{1+0.5x} (2x + y + 1) \, dy \, dx = \int_0^1 \left(2xy + \frac{y^2}{2} + y \right) \Big|_{1-x}^{1+0.5x} dx$$

$$= \int_0^1 \left(\frac{21}{8} x^2 + 3x \right) dx$$

$$= \left(\frac{7}{8} x^3 + \frac{3}{2} x^2 \right) \Big|_0^1 dx$$

$$= \frac{19}{8},$$

and the double integral for R_2:

$$\int_1^2 \int_{2x-2}^{1+0.5x} (2x + y + 1) \, dy \, dx = \int_1^2 (2xy + y^2/2 + y) \Big|_{2x-2}^{1+0.5x} dx$$

$$= \int_0^1 \left(-\frac{39}{8} x^2 + 9x + \frac{3}{2} \right) dx$$

$$= \left(-\frac{13}{8} x^3 + \frac{9}{2} x^2 + \frac{3}{2} x \right) \Big|_1^2$$

$$= \frac{29}{8}.$$

So, Volume $= \dfrac{19}{8} + \dfrac{29}{8} = \dfrac{48}{8} = 6$.

53. Assume the length of the two legs of the right triangle are a and b, respectively. See Figure 16.9. The line through $(a, 0)$ and $(0, b)$ is given by $\frac{y}{b} + \frac{x}{a} = 1$. So the area of this triangle is

$$A = \frac{1}{2} ab.$$

Thus the average distance from the points in the triangle to the y-axis (one of the legs) is

$$\text{Average distance} = \frac{1}{A} \int_0^a \int_0^{-\frac{b}{a}x+b} x \, dy \, dx$$

$$= \frac{2}{ab} \int_0^a \left(-\frac{b}{a} x^2 + bx \right) dx$$

$$= \frac{2}{ab} \left(-\frac{b}{3a} x^3 + \frac{b}{2} x^2 \right) \Big|_0^a$$

$$= \frac{2}{ab} \left(\frac{a^2 b}{6} \right) = \frac{a}{3}.$$

Similarly, the average distance from the points in the triangle to the x-axis (the other leg) is

$$\text{Average distance} = \frac{1}{A} \int_0^b \int_0^{-\frac{a}{b}y+a} y \, dx \, dy$$

$$= \frac{2}{ab} \int_0^b \left(-\frac{a}{b}y^2 + ay\right) dy$$

$$= \frac{2}{ab}\left(\frac{ab^2}{6}\right) = \frac{b}{3}.$$

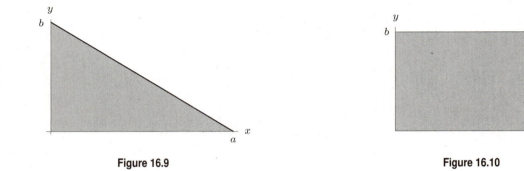

Figure 16.9   **Figure 16.10**

57. The force, ΔF, acting on ΔA, a small piece of area, is given by

$$\Delta F \approx p\Delta A,$$

where p is the pressure at that point. Thus, if R is the rectangle, the total force is given by

$$F = \int_R p \, dA.$$

We choose coordinates with the origin at one corner of the plate. See Figure 16.10. Suppose p is proportional to the square of the distance from the corner represented by the origin. Then we have

$$p = k(x^2 + y^2), \quad \text{for some positive constant } k.$$

Thus, we want to compute $\int_R k(x^2 + y^2)dA$. Rewriting as an iterated integral, we have

$$F = \int_R k(x^2 + y^2)\, dA = \int_0^b \int_0^a k(x^2 + y^2)\, dxdy = k\int_0^b \left(\frac{x^3}{3} + xy^2 \Big|_0^a\right) dy$$

$$= k\int_0^b \left(\frac{a^3}{3} + ay^2\right) dy = k\left(\frac{a^3 y}{3} + a\frac{y^3}{3}\Big|_0^b\right)$$

$$= \frac{k}{3}(a^3 b + ab^3).$$

Solutions for Section 16.3

Exercises

1.

$$\int_W f\, dV = \int_0^2 \int_{-1}^1 \int_2^3 (x^2 + 5y^2 - z)\, dz\, dy\, dx$$

$$= \int_0^2 \int_{-1}^1 (x^2 z + 5y^2 z - \frac{1}{2}z^2)\Big|_2^3 dy\, dx$$

$$= \int_0^2 \int_{-1}^1 (x^2 + 5y^2 - \frac{5}{2})\, dy\, dx$$

$$= \int_0^2 (x^2 y + \frac{5}{3}y^3 - \frac{5}{2}y)\Big|_{-1}^1 dx$$

$$= \int_0^2 (2x^2 + \frac{10}{3} - 5)\, dx$$

$$= (\frac{2}{3}x^3 - \frac{5}{3}x)\Big|_0^2$$

$$= \frac{16}{3} - \frac{10}{3} = 2$$

5. The region is the half cylinder in Figure 16.11.

9. The region is the cylinder in Figure 16.12.

13. The region is the quarter sphere in Figure 16.13.

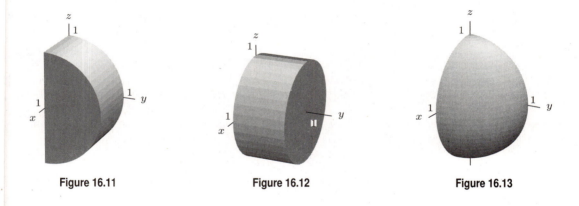

| Figure 16.11 | Figure 16.12 | Figure 16.13 |

Problems

17. The sphere $x^2 + y^2 + z^2 = 9$ intersects the plane $z = 2$ in the circle

$$x^2 + y^2 + 2^2 = 9$$
$$x^2 + y^2 = 5.$$

The upper half of the sphere is given by $z = \sqrt{9 - x^2 - y^2}$. Thus, using the limits from Figure 16.14 gives

$$V = \int_{-\sqrt{5}}^{\sqrt{5}} \int_{-\sqrt{5-x^2}}^{\sqrt{5-x^2}} \int_2^{\sqrt{9-x^2-y^2}} 1\, dz\, dy\, dx.$$

The order of integration of x and y can be reversed.

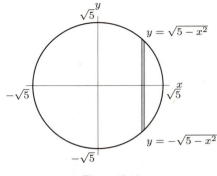

Figure 16.14

21. A slice through W for a fixed value of x is a semi-circle the boundary of which is $z^2 = r^2 - x^2 - y^2$, for $z \geq 0$, so the inner integral is

$$\int_0^{\sqrt{r^2-x^2-y^2}} f(x, y, z) \, dz.$$

Lining up these stacks parallel to y-axis gives a slice from $y = -\sqrt{r^2 - x^2}$ to $y = \sqrt{r^2 - x^2}$ giving

$$\int_{-\sqrt{r^2-x^2}}^{\sqrt{r^2-x^2}} \int_0^{\sqrt{r^2-x^2-y^2}} f(x, y, z) \, dz \, dy.$$

Finally, there is a slice for each x between 0 and r, so the integral we want is

$$\int_0^r \int_{-\sqrt{r^2-x^2}}^{\sqrt{r^2-x^2}} \int_0^{\sqrt{r^2-x^2-y^2}} f(x, y, z) \, dz \, dy \, dx.$$

25. The pyramid is shown in Figure 16.15. The planes $y = 0$, and $y - x = 4$, and $2x + y + z = 4$ intersect the plane $z = -6$ in the lines $y = 0$, $y - x = 4$, $2x + y = 10$ on the $z = -6$ plane as shown in Figure 16.16.

These three lines intersect at the points $(-4, 0, -6)$, $(5, 0, -6)$, and $(2, 6, -6)$. Let R be the triangle in the planes $z = -6$ with the above three points as vertices. Then, the volume of the solid is

$$V = \int_0^6 \int_{y-4}^{(10-y)/2} \int_{-6}^{4-2x-y} dz \, dx \, dy$$

$$= \int_0^6 \int_{y-4}^{(10-y)/2} (10 - 2x - y) \, dx \, dy = 162$$

$$= \int_0^6 (10x - x^2 - xy) \Big|_{y-4}^{(10-y)/2} dy$$

$$= \int_0^6 (\frac{9y^2}{4} - 27y + 81) \, dy \qquad = \frac{9y^3}{12} - \frac{27y^2}{2} + 81y$$

$$= 162$$

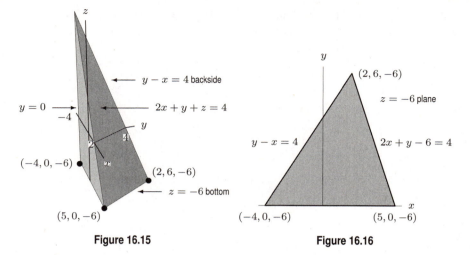

Figure 16.15   **Figure 16.16**

29. The required volume, V, is given by

$$V = \int_0^5 \int_0^3 \int_0^{x^2} dz\, dy\, dx$$

$$= \int_0^5 \int_0^3 x^2\, dy\, dx$$

$$= \int_0^5 x^2 y \Big|_{y=0}^{y=3} dx$$

$$= \int_0^5 3x^2\, dx$$

$$= 125.$$

33. (a) The vectors $\vec{u} = \vec{i} - \vec{j}$ and $\vec{v} = \vec{i} - \vec{k}$ lie in the required plane so $\vec{p} = \vec{u} \times \vec{v} = \vec{i} + \vec{j} + \vec{k}$ is perpendicular to this plane. Let (x, y, z) be a point in the plane, then $(x-1)\vec{i} + y\vec{j} + z\vec{k}$ is perpendicular to $\vec{p}$, so $((x-1)\vec{i} + y\vec{j} + z\vec{k}) \cdot (\vec{i} + \vec{j} + \vec{j}) = 0$ and so

$$(x-1) + y + z = 0.$$

Therefore, the equation of the required plane is $x + y + z = 1$.

(b) The required volume, V, is given by

$$V = \int_0^1 \int_0^{1-x} \int_0^{1-x-y} dz\, dy\, dx$$

$$= \int_0^1 \int_0^{1-x} (1 - x - y)\, dy\, dx$$

$$= \int_0^1 y - xy - \frac{1}{2}y^2 \Big|_0^{1-x} dx$$

$$= \int_0^1 \left(1 - x - x(1-x) - \frac{1}{2}(1-x)^2\right) dx$$

$$= \int_0^1 \frac{1}{2}(1-x)^2 dx$$

$$= \frac{1}{6}.$$

37. From the problem, we know that (x, y, z) is in the cube which is bounded by the three coordinate planes, $x = 0$, $y = 0$, $z = 0$ and the planes $x = 2$, $y = 2$, $z = 2$. We can regard the value $x^2 + y^2 + z^2$ as the density of the cube. The average value of $x^2 + y^2 + z^2$ is given by

$$\text{average value} = \frac{\int_V (x^2 + y^2 + z^2)\, dV}{\text{volume}(V)}$$

$$= \frac{\int_0^2 \int_0^2 \int_0^2 (x^2 + y^2 + z^2)\, dx dy dz}{8}$$

$$= \frac{\int_0^2 \int_0^2 \left(\frac{x^3}{3} + (y^2 + z^2)x\right) \Big|_0^2 dy dz}{8}$$

$$= \frac{\int_0^2 \int_0^2 \left(\frac{8}{3} + 2y^2 + 2z^2\right) dy dz}{8}$$

$$= \frac{\int_0^2 \left(\frac{8}{3}y + \frac{2}{3}y^3 + 2z^2 y\right) \Big|_0^2 dz}{8}$$

$$= \frac{\int_0^2 \left(\frac{16}{3} + \frac{16}{3} + 4z^2\right) dz}{8}$$

$$= \frac{\left(\frac{32}{3}z + \frac{4}{3}z^3\right) \Big|_0^2}{8}$$

$$= \frac{\left(\frac{64}{3} + \frac{32}{3}\right)}{8} = 4.$$

41. Zero. The value of y is positive on the half of the cone above the second and third quadrants and negative (of equal absolute value) on the half of the cone above the third and fourth quadrants. The integral of y over the entire solid cone is zero because the integrals over the four quadrants cancel.

45. Negative. If (x, y, z) is any point inside the cone then $z < 2$. Hence the function $z - 2$ is negative on W and so is its integral.

49. Positive. The value of x is positive on the half-cone, so its integral is positive.

53. Zero. Write the triple integral as an iterated integral, say integrating first with respect to y. For fixed x and z, the y-integral is over an interval symmetric about 0. The integral of y over such an interval is zero. If any of the inner integrals in an iterated integral is zero, then the triple integral is zero.

57. The intersection of two cylinders $x^2 + z^2 = 1$ and $y^2 + z^2 = 1$ is shown in Figure 16.17. This region is bounded by four surfaces:

$$z = -\sqrt{1 - x^2}, \quad z = \sqrt{1 - x^2}, \quad y = -\sqrt{1 - z^2}, \quad \text{and} \quad y = \sqrt{1 - z^2}$$

So the volume of the given solid is

$$V = \int_{-1}^{1} \int_{-\sqrt{1-x^2}}^{\sqrt{1-x^2}} \int_{-\sqrt{1-z^2}}^{\sqrt{1-z^2}} dy\, dz\, dx$$

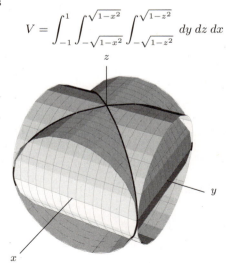

Figure 16.17

61. The volume V of the solid is $1 \cdot 2 \cdot 3 = 6$. We need to compute

$$
\begin{aligned}
\frac{m}{6} \int_W x^2 + y^2 \, dV &= \frac{m}{6} \int_0^1 \int_0^2 \int_0^3 x^2 + y^2 \, dz \, dy \, dx \\
&= \frac{m}{6} \int_0^1 \int_0^2 3(x^2 + y^2) \, dy \, dx \\
&= \frac{m}{2} \int_0^1 (x^2 y + y^3/3)\Big|_0^2 \, dx \\
&= \frac{m}{2} \int_0^1 (2x^2 + 8/3) \, dx = 5m/3
\end{aligned}
$$

Solutions for Section 16.4

Exercises

1. $\displaystyle \int_0^{2\pi} \int_0^{\sqrt{2}} f \, r \, dr \, d\theta$

5. A circle is best described in polar coordinates. The radius is 5, so r goes from 0 to 5. To include the entire circle, we need θ to go from 0 to 2π. The integral is

$$
\int_0^{2\pi} \int_0^5 f(r \cos\theta, r \sin\theta) \, r \, dr \, d\theta.
$$

9. See Figure 16.18.

13. See Figure 16.19.

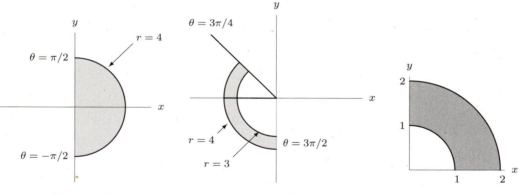

Figure 16.18   **Figure 16.19**   **Figure 16.20**

17. The region is pictured in Figure 16.20. Using polar coordinates, we get

$$
\begin{aligned}
\int_R (x^2 - y^2) dA &= \int_0^{\pi/2} \int_1^2 r^2(\cos^2\theta - \sin^2\theta) r \, dr \, d\theta = \int_0^{\pi/2} (\cos^2\theta - \sin^2\theta) \cdot \frac{1}{4} r^4 \Big|_1^2 \, d\theta \\
&= \frac{15}{4} \int_0^{\pi/2} (\cos^2\theta - \sin^2\theta) \, d\theta \\
&= \frac{15}{4} \int_0^{\pi/2} \cos 2\theta \, d\theta \\
&= \frac{15}{4} \cdot \frac{1}{2} \sin 2\theta \Big|_0^{\pi/2} = 0.
\end{aligned}
$$

Problems

21. From the given limits, the region of integration is in Figure 16.21.

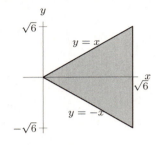

Figure 16.21

In polar coordinates, $-\pi/4 \le \theta \le \pi/4$. Also, $\sqrt{6} = x = r\cos\theta$. Hence, $0 \le r \le \sqrt{6}/\cos\theta$. The integral becomes

$$\int_0^{\sqrt{6}} \int_{-x}^{x} dy\,dx = \int_{-\pi/4}^{\pi/4} \int_0^{\sqrt{6}/\cos\theta} r\,dr\,d\theta$$

$$= \int_{-\pi/4}^{\pi/4} \left(\frac{r^2}{2} \Big|_0^{\sqrt{6}/\cos\theta} \right) d\theta = \int_{-\pi/4}^{\pi/4} \frac{6}{2\cos^2\theta}\,d\theta$$

$$= 3\tan\theta \Big|_{-\pi/4}^{\pi/4} = 3\cdot(1-(-1)) = 6.$$

Notice that we can check this answer because the integral gives the area of the shaded triangular region which is $\frac{1}{2}\cdot\sqrt{6}\cdot(2\sqrt{6}) = 6$.

25. The average value of the function r on the disc R of radius a is

$$\text{Average of } r = \frac{1}{\text{Area of } R} \int_R r\,dA = \frac{1}{\pi a^2} \int_0^{2\pi} \int_0^a rr\,dr\,d\theta = \frac{1}{\pi a^2} \int_0^{2\pi} \frac{a^3}{3}\,d\theta = \frac{1}{\pi a^2} 2\pi \frac{a^3}{3} = \frac{2a}{3}.$$

29. A rough graph of the base of the spring is in Figure 16.22, where the coil is roughly of width 0.01 inches. The volume is equal to the product of the base area and the height. To calculate the area we use polar coordinates, taking the following integral:

$$\text{Area} = \int_0^{4\pi} \int_{0.25+0.04\theta}^{0.26+0.04\theta} r\,dr\,d\theta$$

$$= \frac{1}{2} \int_0^{4\pi} (0.26 - 0.04\theta)^2 - (0.25 - 0.04\theta)^2\,d\theta$$

$$= \frac{1}{2} \int_0^{4\pi} 0.01\cdot(0.51 + 0.08\theta)\,d\theta$$

$$= 0.0051\cdot 2\pi + \frac{1}{4}(0.0008\theta^2) \Big|_0^{4\pi} = 0.0636$$

Therefore, the volume$= 0.0636\cdot 0.2 = 0.0127$ in^3.

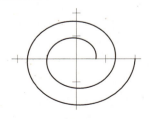

Figure 16.22

33. (a) The curve $r = 1/(2\cos\theta)$ or $r\cos\theta = 1/2$ is the line $x = 1/2$. The curve $r = 1$ is the circle of radius 1 centered at the origin. See Figure 16.23.

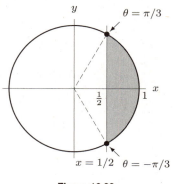

Figure 16.23

(b) The line intersects the circle where $2\cos\theta = 1$, so $\theta = \pm\pi/3$. From Figure 16.23 we see that

$$\text{Area} = \int_{-\pi/3}^{\pi/3}\int_{1/(2\cos\theta)}^{1} r\,dr\,d\theta.$$

Evaluating gives

$$\text{Area} = \int_{-\pi/3}^{\pi/3}\left(\frac{r^2}{2}\bigg|_{1/(2\cos\theta)}^{1}\right)d\theta = \frac{1}{2}\int_{-\pi/3}^{\pi/3}\left(1 - \frac{1}{4\cos^2\theta}\right)d\theta$$

$$= \frac{1}{2}\left(\theta - \frac{\tan\theta}{4}\right)\bigg|_{-\pi/3}^{\pi/3} = \frac{1}{2}\left(2\frac{\pi}{3} - 2\frac{\sqrt{3}}{4}\right) = \frac{4\pi - 3\sqrt{3}}{12}.$$

Solutions for Section 16.5

Exercises

1. (a) is (IV); (b) is (II); (c) is (VII); (d) is (VI); (e) is (III); (f) is (V).

5. The plane has equation $\theta = \pi/4$.

9.

$$\int_W f\,dV = \int_{-1}^{3}\int_{0}^{2\pi}\int_{0}^{1}(\sin(r^2))\,r\,dr\,d\theta\,dz$$

$$= \int_{-1}^{3}\int_{0}^{2\pi}\left(-\frac{1}{2}\cos r^2\right)\bigg|_{0}^{1}d\theta\,dz$$

$$= -\frac{1}{2}\int_{-1}^{3}\int_{0}^{2\pi}(\cos 1 - \cos 0)\,d\theta\,dz$$

$$= -\pi\int_{-1}^{3}(\cos 1 - 1)\,dz = -4\pi(\cos 1 - 1) = 4\pi(1 - \cos 1)$$

13. Using cylindrical coordinates, we get:

$$\int_0^1 \int_0^{2\pi} \int_0^4 f \cdot r\, dr\, d\theta\, dz$$

17. We use Cartesian coordinates, oriented as shown in Figure 16.24. The slanted top has equation $z = mx$, where m is the slope in the x-direction, so $m = 1/5$. Then if f is an arbitrary function, the triple integral is

$$\int_0^5 \int_0^2 \int_0^{x/5} f\, dz\, dy\, dx.$$

Other answers are possible.

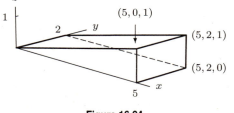

Figure 16.24

Problems

21. We want the volume of the region above the cone $\phi = \pi/3$ and below the sphere $\rho = 3$:

$$V = \int_0^{2\pi} \int_0^{\pi/3} \int_0^3 \rho^2 \sin\phi\, d\rho\, d\phi\, d\theta.$$

The order of integration can be altered and other coordinates can be used.

25. We use cylindrical coordinates since the sphere $x^2 + y^2 + z^2 = 10$, or $r^2 + z^2 = 10$, and the plane $z = 1$ can both be simply expressed. The plane cuts the sphere in the circle $r^2 + 1^2 = 10$, or $r = 3$. Thus

$$V = \int_0^{2\pi} \int_0^3 \int_1^{\sqrt{10-r^2}} r\, dz\, dr\, d\theta,$$

or

$$V = \int_0^{2\pi} \int_1^{\sqrt{10}} \int_0^{\sqrt{10-z^2}} r\, dr\, dz\, d\theta.$$

Order of integration can be altered and other coordinates can be used.

29. In rectangular coordinates, a cone has equation $z = k\sqrt{x^2 + y^2}$ for some constant k. Since $z = 4$ when $\sqrt{x^2 + y^2} = \sqrt{2^2} = 2$, we have $k = 2$. Thus, the integral is

$$\int_{-2}^2 \int_{-\sqrt{4-x^2}}^{\sqrt{4-x^2}} \int_{2\sqrt{x^2+y^2}}^4 h(x, y, z)\, dz\, dy\, dx.$$

33. The region is a solid cylinder of height 1, radius 1 with base on the xy-plane and axis on the z-axis. We have:

$$\int_0^1 \int_{-1}^1 \int_{-\sqrt{1-x^2}}^{\sqrt{1-x^2}} \frac{1}{(x^2 + y^2)^{1/2}}\, dy\, dx\, dz = \int_0^1 \int_0^{2\pi} \int_0^1 \frac{1}{r} r\, dr\, d\theta\, dz$$

$$= \int_0^1 \int_0^{2\pi} r \Big|_0^1 d\theta\, dz$$

$$= \int_0^1 \int_0^{2\pi} d\theta\, dz = 2\pi.$$

Note that the integral is improper, but it can be shown that the result is correct.

37. The cone is centered along the positive x-axis and intersects the sphere in the circle

$$(y^2 + z^2) + y^2 + z^2 = 4$$
$$y^2 + z^2 = 2.$$

We use spherical coordinates with ϕ measured from the x-axis and θ measured in the yz-plane. (Alternatively, the volume we want is equal to the volume between the cone $z = \sqrt{x^2 + y^2}$ and the sphere $x^2 + y^2 + z^2 = 4$.) The cone is given by $\phi = \pi/4$. The sphere has equation $\rho = 2$. Thus

$$
\begin{aligned}
\text{Volume} &= \int_0^{2\pi} \int_0^{\pi/4} \int_0^2 \rho^2 \sin\phi \, d\rho \, d\phi \, d\theta \\
&= \int_0^{2\pi} \int_0^{\pi/4} \frac{\rho^3}{3} \sin\phi \Big|_0^2 \, d\phi \, d\theta \\
&= \int_0^{2\pi} \int_0^{\pi/4} \frac{8}{3} \sin\phi \, d\phi \, d\theta \\
&= \int_0^{2\pi} -\frac{8}{3} \cos\phi \Big|_0^{\pi/4} \, d\theta = \int_0^{2\pi} \frac{8}{3}\left(1 - \frac{1}{\sqrt{2}}\right) \, d\theta \\
&= \frac{16\pi}{3}\left(1 - \frac{1}{\sqrt{2}}\right).
\end{aligned}
$$

41. The region whose volume we want is shown in Figure 16.25:

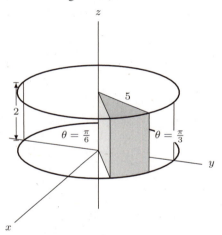

Figure 16.25

Using cylindrical coordinates, the volume is given by the integral:

$$
\begin{aligned}
V &= \int_0^2 \int_{\pi/6}^{\pi/3} \int_0^5 r \, dr \, d\theta \, dz \\
&= \int_0^2 \int_{\pi/6}^{\pi/3} \frac{r^2}{2}\Big|_0^5 \, d\theta \, dz \\
&= \frac{25}{2} \int_0^2 \int_{\pi/6}^{\pi/3} d\theta \, dz \\
&= \frac{25}{2} \int_0^2 \left(\frac{\pi}{3} - \frac{\pi}{6}\right) \, dz \\
&= \frac{25}{2} \cdot \frac{\pi}{6} \cdot 2 = \frac{25\pi}{6}.
\end{aligned}
$$

45. The cylinder has radius 2. Using cylindrical coordinates to find the mass and integrating with respect to r first, we have

$$\text{Mass} = \int_0^{2\pi} \int_0^3 \int_0^2 (1+r) r \, dr \, dz \, d\theta = \int_0^{2\pi} \int_0^3 \left(\frac{r^2}{2} + \frac{r^3}{3} \right) \Bigg|_0^2 dz \, d\theta = 2\pi \cdot 3 \cdot \left(\frac{4}{2} + \frac{8}{3} \right) = 28\pi \text{ gm.}$$

49. (a) We use spherical coordinates. Since $\delta = 9$ where $\rho = 6$ and $\delta = 11$ where $\rho = 7$, the density increases at a rate of 2 gm/cm^3 for each cm increase in radius. Thus, since density is a linear function of radius, the slope of the linear function is 2. Its equation is

$$\delta - 11 = 2(\rho - 7) \quad \text{so} \quad \delta = 2\rho - 3.$$

(b) Thus,

$$\text{Mass} = \int_0^{2\pi} \int_0^\pi \int_6^7 (2\rho - 3)\rho^2 \sin\phi \, d\rho \, d\phi \, d\theta.$$

(c) Evaluating the integral, we have

$$\text{Mass} = 2\pi \left(-\cos\phi \Big|_0^\pi \right) \left(\frac{2\rho^4}{4} - \frac{3\rho^3}{3} \Big|_6^7 \right) = 2\pi \cdot 2(425.5) = 1702\pi \text{ gm} = 5346.991 \text{ gm.}$$

53. We must first decide on coordinates. We pick spherical coordinates with the common center of the two spheres as the origin. We imagine the half-melon with the flat side horizontal and the positive z-axis going through the curved surface. See Figure 16.26. The volume is given by the integral

$$\text{Volume} = \int_0^{2\pi} \int_0^{\pi/2} \int_a^b \rho^2 \sin\phi \, d\rho \, d\phi \, d\theta.$$

Evaluating gives

$$\text{Volume} = \int_0^{2\pi} \int_0^{\pi/2} \sin\phi \, \frac{\rho^3}{3} \Big|_{\rho=a}^{\rho=b} d\phi \, d\theta = 2\pi(-\cos\phi) \Big|_0^{\pi/2} \left(\frac{b^3}{3} - \frac{a^3}{3} \right) = \frac{2\pi}{3}(b^3 - a^3).$$

To check our answer, notice that the volume is the difference between the volumes of two half spheres of radius a and b. These half spheres have volumes $2\pi b^3/3$ and $2\pi a^3/3$, respectively.

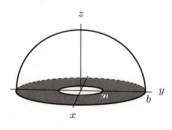

Figure 16.26

57. We first need to find the mass of the solid, using cylindrical coordinates:

$$m = \int_0^{2\pi} \int_0^1 \int_0^{\sqrt{z/a}} r \, dr \, dz \, d\theta$$

$$= \int_0^{2\pi} \int_0^1 \frac{z}{2a} \, dz \, d\theta$$

$$= \int_0^{2\pi} \frac{1}{4a} \, d\theta = \frac{\pi}{2a}$$

It makes sense that the mass would vary inversely with a, since increasing a makes the paraboloid skinnier. Now for the z-coordinate of the center of mass, again using cylindrical coordinates:

$$\bar{z} = \frac{2a}{\pi} \int_0^{2\pi} \int_0^1 \int_0^{\sqrt{z/a}} zr \, dr \, dz \, d\theta$$

$$= \frac{2a}{\pi} \int_0^{2\pi} \int_0^1 \frac{z^2}{2a} \, dz \, d\theta$$

$$= \frac{2a}{\pi} \int_0^{2\pi} \frac{1}{6a} \, d\theta = \frac{2}{3}$$

61. Assume the base of the cylinder sits on the xy-plane with center at the origin. Because the cylinder is symmetric about the z-axis, the force in the horizontal x or y direction is 0. Thus we need only compute the vertical z component of the force. We are going to use cylindrical coordinates; since the force is $G \cdot \text{mass}/(\text{distance})^2$, a piece of the cylinder of volume dV located at (r, θ, z) exerts on the unit mass a force with magnitude $G(\delta \, dV)/(r^2 + z^2)$. See Figure 16.27.

$$\begin{array}{c} \text{Vertical component} \\ \text{of force} \end{array} = \frac{G(\delta \, dV)}{r^2 + z^2} \cdot \cos\phi = \frac{G\delta \, dV}{r^2 + z^2} \cdot \frac{z}{\sqrt{r^2 + z^2}} = \frac{G\delta z \, dV}{(r^2 + z^2)^{3/2}}.$$

Adding up all the contributions of all the dV's, we obtain

$$\text{Vertical force} = \int_0^H \int_0^{2\pi} \int_0^R \frac{G\delta zr}{(r^2 + z^2)^{3/2}} \, dr \, d\theta \, dz$$

$$= \int_0^H \int_0^{2\pi} (G\delta z) \left(-\frac{1}{\sqrt{r^2 + z^2}} \right) \Big|_0^R \, d\theta \, dz$$

$$= \int_0^H \int_0^{2\pi} (G\delta z) \cdot \left(-\frac{1}{\sqrt{R^2 + z^2}} + \frac{1}{z} \right) \, d\theta \, dz$$

$$= \int_0^H 2\pi G\delta \left(1 - \frac{z}{\sqrt{R^2 + z^2}} \right) \, dz$$

$$= 2\pi G\delta (z - \sqrt{R^2 + z^2}) \Big|_0^H$$

$$= 2\pi G\delta (H - \sqrt{R^2 + H^2} + R) = 2\pi G\delta (H + R - \sqrt{R^2 + H^2})$$

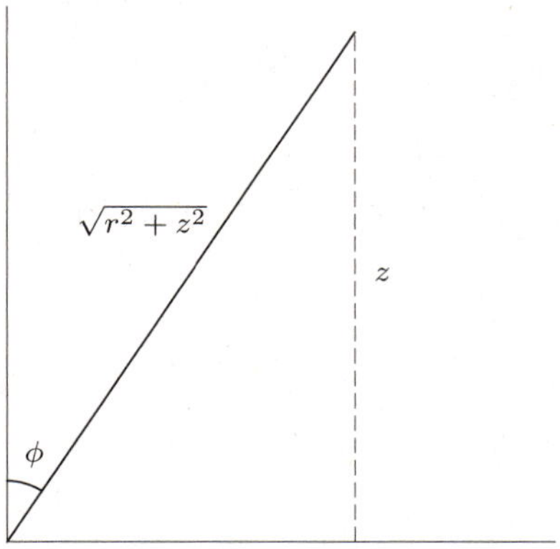

Figure 16.27

65. Use cylindrical coordinates, with the z-axis being the axis of the cable. Consider a piece of cable of length 1. Then

$$\text{Stored energy} = \frac{1}{2} \int_a^b \int_0^1 \int_0^{2\pi} \epsilon E^2 \, r \, d\theta \, dz \, dr = \frac{q^2}{8\pi^2\epsilon} \int_a^b \int_0^1 \int_0^{2\pi} \frac{1}{r} \, d\theta \, dz \, dr$$

$$= \frac{q^2}{4\pi\epsilon} \int_a^b \frac{1}{r} \, dr = \frac{q^2}{4\pi\epsilon} (\ln b - \ln a) = \frac{q^2}{4\pi\epsilon} \ln \frac{b}{a}.$$

So the stored energy is proportional to $\ln(b/a)$ with constant of proportionality $q^2/4\pi\epsilon$.

Solutions for Section 16.6

Exercises

1. We have $p(x, y) = 0$ for all points (x, y) satisfying $x \geq 3$, since all such points lie outside the region R. Therefore the fraction of the population satisfying $x \geq 3$ is 0.

5. Since $p(x, y) = 0$ for all (x, y) outside the rectangle R, the population is given by the volume under the graph of p over the region inside the rectangle R and to the right of the line $x = y$. Therefore the fraction of the population is given by the double integral:

$$\int_0^1 \int_y^2 xy\, dx\, dy = \int_0^1 \frac{x^2 y}{2}\Big|_y^2 dy = \int_0^1 \left(2y - \frac{y^3}{2}\right) dy = \left(y^2 - \frac{y^4}{8}\right)\Big|_0^1 = \frac{7}{8}.$$

9. No, p is not a joint density function. Since $p(x, y) = 0$ outside the region R, the volume under the graph of p is the same as the volume under the graph of p over the region R, which is 2 not 1.

13. Yes, p is a joint density function. In the region R we have $1 \geq x^2 + y^2$, so $p(x, y) = (2/\pi)(1 - x^2 - y^2) \geq 0$ for all x and y in R, and $p(x, y) = 0$ for all other (x, y). To check that p is a joint density function, we check that the total volume under the graph of p over the region R is 1. Using polar coordinates, we get:

$$\int_R p(x, y)dA = \frac{2}{\pi}\int_0^{2\pi}\int_0^1 (1 - r^2)r\, dr\, d\theta = \frac{2}{\pi}\int_0^{2\pi}\left(\frac{r^2}{2} - \frac{r^4}{4}\right)\Big|_0^1 d\theta = \frac{2}{\pi}\int_0^{2\pi}\frac{1}{4} d\theta = 1.$$

Problems

17. (a) For a density function,

$$1 = \int_{-\infty}^{\infty}\int_{-\infty}^{\infty} f(x, y)\, dy\, dx = \int_0^2\int_0^1 kx^2\, dy\, dx$$

$$= \int_0^2 kx^2\, dx$$

$$= \frac{kx^3}{3}\Big|_0^2 = \frac{8k}{3}.$$

So $k = 3/8$.

(b)
$$\int_0^1\int_0^{2-y} \frac{3}{8}x^2\, dx\, dy = \int_0^1 \frac{1}{8}(2-y)^3\, dy = \frac{-1}{32}(2-y)^4\Big|_0^1 = \frac{15}{32}$$

(c)
$$\int_0^{1/2}\int_0^1 \frac{3}{8}x^2\, dx\, dy = \int_0^{1/2}\frac{1}{8}x^3\Big|_0^1 dy = \int_0^{1/2}\frac{1}{8}\, dy = \frac{1}{16}.$$

21. (a) If $t \leq 0$, then $F(t) = 0$ because the average of two positive numbers can not be negative. If $1 < t$ then $F(t) = 1$ because the average of two numbers each at most 1 is certain to be less than or equal to 1. For any t, we have $F(t) = \int_R p(x, y)dA$ where R is the region of the plane defined by $(x + y)/2 \leq t$. Since $p(x, y) = 0$ outside the unit square, we need integrate only over the part of R that lies inside the square, and since $p(x, y) = 1$ inside the square, the integral equals the area of that part of the square. Thus, we can calculate the area using area formulas. For $0 \leq t \leq 1$, we draw the line $(x + y)/2 = t$, which has x- and y-intercepts of $2t$. Figure 16.28 shows that for $0 < t \leq 1/2$,

$$F(t) = \text{ Area of triangle } = \frac{1}{2}\cdot 2t \cdot 2t = 2t^2.$$

In Figure 16.29, when $x = 1$, we have $y = 2t - 1$. Thus, the vertical side of the unshaded triangle is $1 - (2t - 1) = 2 - 2t$. The horizontal side is the same length, so for $1/2 < t \leq 1$,

$$F(t) = \text{ Area of Square } - \text{ Area of triangle } = 1^2 - \frac{1}{2}(2 - 2t)^2 - 1 - 2(1 - t)^2.$$

The final result is:

$$F(t) = \begin{cases} 0 & \text{if } t \leq 0 \\ 2t^2 & \text{if } 0 < t \leq 1/2 \\ 1 - 2(1-t)^2 & \text{if } 1/2 < t \leq 1 \\ 1 & \text{if } 1 < t \end{cases}.$$

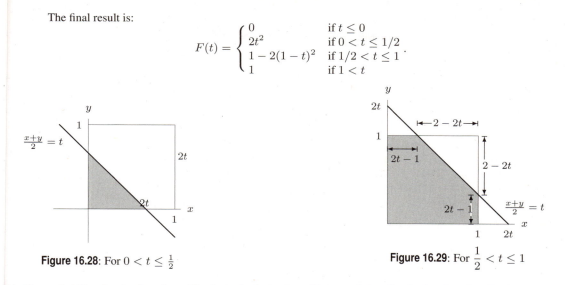

Figure 16.28: For $0 < t \leq \frac{1}{2}$

Figure 16.29: For $\frac{1}{2} < t \leq 1$

(b) The probability density function $p(t)$ of z is the derivative of its cumulative distribution function. We have

$$p(t) = \begin{cases} 0 & \text{if } t \leq 0 \\ 4t & \text{if } 0 < t \leq 1/2 \\ 4 - 4t & \text{if } 1/2 < t \leq 1 \\ 0 & \text{if } 1 < t \end{cases}.$$

See Figure 16.30.

(c) The values of x and y and equally likely to be near $0, 1/2$, and 1. Notice from the graph of the density function in Figure 16.30 that even though x and y separately are equally likely to be anywhere between 0 and 1, their average $z = (x + y)/2$ is more likely to be near $1/2$ than to be near 0 or 1.

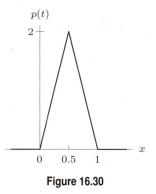

Figure 16.30

Solutions for Section 16.7

Exercises

1. We have

$$\frac{\partial(x, y)}{\partial(s, t)} = \begin{vmatrix} x_s & x_t \\ y_s & y_t \end{vmatrix} = \begin{vmatrix} 5 & 2 \\ 3 & 1 \end{vmatrix} = -1.$$

Therefore,

$$\left| \frac{\partial(x, y)}{\partial(s, t)} \right| = 1.$$

5. We have

$$\frac{\partial(x, y, z)}{\partial(s, t, u)} = \begin{vmatrix} x_s & x_t & x_u \\ y_s & y_t & y_u \\ z_s & z_t & z_u \end{vmatrix} = \begin{vmatrix} 3 & 1 & 2 \\ 1 & 5 & -1 \\ 2 & -1 & 1 \end{vmatrix}.$$

This 3×3 determinant is computed the same way as for the cross product, with the entries $3, 1, 2$ in the first row playing the same role as $\vec{i}, \vec{j}, \vec{k}$. We get

$$\frac{\partial(x, y, z)}{\partial(s, t, u)} = ((5)(1) - (-1)(-1))3 + ((-1)(2) - (1)(1))1 + ((1)(-1) - (2)(5))2 = -13.$$

9. The square T is defined by the inequalities

$$0 \le s = ax \le 1 \qquad 0 \le t = by \le 1$$

that correspond to the inequalities

$$0 \le x \le 1/a = 50 \qquad 0 \le y \le 1/b = 10$$

that define R. Thus $a = 1/50$ and $b = 1/10$.

13. Inverting the change of variables gives $x = s - at, y = t$.

The four edges of R are

$$y = 0, y = 5, y = -\frac{1}{3}x, y = -\frac{1}{3}(x - 10).$$

The change of variables transforms the edges to

$$t = 0, t = 5, t = -\frac{1}{3}s + \frac{1}{3}at, t = -\frac{1}{3}s + \frac{1}{3}at + \frac{10}{3}.$$

These are equations for the edges of a rectangle in the st-plane if the last two are of the form: $s = $ (Constant). This happens when the t terms drop out, or $a = 3$. With $a = 3$ the change of variables gives

$$\int \int_T \left| \frac{\partial(x, y)}{\partial(s, t)} \right| ds \, dt$$

over the rectangle

$$T : 0 \le t \le 5, 0 \le s \le 10.$$

Problems

17. Given

$$\begin{cases} x = \rho \sin \phi \cos \theta \\ y = \rho \sin \phi \sin \theta \\ z = \rho \cos \phi, \end{cases}$$

$$\frac{\partial(x, y, z)}{\partial(\rho, \phi, \theta)} = \begin{vmatrix} \frac{\partial x}{\partial \rho} & \frac{\partial x}{\partial \phi} & \frac{\partial x}{\partial \theta} \\ \frac{\partial y}{\partial \rho} & \frac{\partial y}{\partial \phi} & \frac{\partial y}{\partial \theta} \\ \frac{\partial z}{\partial \rho} & \frac{\partial z}{\partial \phi} & \frac{\partial z}{\partial \theta} \end{vmatrix} = \begin{vmatrix} \sin \phi \cos \theta & \rho \cos \phi \cos \theta & -\rho \sin \phi \sin \theta \\ \sin \phi \sin \theta & \rho \cos \phi \sin \theta & \rho \sin \phi \cos \theta \\ \cos \phi & -\rho \sin \phi & 0 \end{vmatrix}$$

$$= \cos \phi \begin{vmatrix} \rho \cos \phi \cos \theta & -\rho \sin \phi \sin \theta \\ \rho \cos \phi \sin \theta & \rho \sin \phi \cos \theta \end{vmatrix} + \rho \sin \phi \begin{vmatrix} \sin \phi \cos \theta & -\rho \sin \phi \sin \theta \\ \sin \phi \sin \theta & \rho \sin \phi \cos \theta \end{vmatrix}$$

$$= \cos \phi (\rho^2 \cos^2 \theta \cos \phi \sin \phi + \rho^2 \sin^2 \theta \cos \phi \sin \phi)$$

$$\quad + \rho \sin \phi (\rho \sin^2 \phi \cos^2 \theta + \rho \sin^2 \phi \sin^2 \theta)$$

$$= \rho^2 \cos^2 \phi \sin \phi + \rho^2 \sin^3 \phi$$

$$= \rho^2 \sin \phi.$$

21. The area of the ellipse is $\int\int_R dx\,dy$ where R is the region $x^2 + 2xy + 2y^2 \leq 1$. We must change variables in both the area element $dA = dx\,dy$ and the region R.

Inverting the variable change gives $x = s - t$, $y = t$. Thus

$$\frac{\partial(x,y)}{\partial(s,t)} = \begin{vmatrix} \frac{\partial x}{\partial s} & \frac{\partial x}{\partial t} \\ \frac{\partial y}{\partial s} & \frac{\partial y}{\partial t} \end{vmatrix} = \begin{vmatrix} 1 & -1 \\ 0 & 1 \end{vmatrix} = 1.$$

Therefore

$$dx\,dy = \left| \frac{\partial(x,y)}{\partial(s,t)} \right| ds\,dt = ds\,dt.$$

The region of integration is

$$x^2 + 2xy + 2y^2 = (s-t)^2 + 2(s-t)t + 2t^2 = s^2 + t^2 \leq 1.$$

Let T be the unit disc $s^2 + t^2 \leq 1$. We have

$$\int\int_R dx\,dy = \int\int_T ds\,dt = \text{Area of } T = \pi.$$

25. Given

$$\begin{cases} s = xy \\ t = xy^2, \end{cases}$$

we have

$$\frac{\partial(s,t)}{\partial(x,y)} = \begin{vmatrix} \frac{\partial s}{\partial x} & \frac{\partial s}{\partial y} \\ \frac{\partial t}{\partial x} & \frac{\partial t}{\partial y} \end{vmatrix} = \begin{vmatrix} y & x \\ y^2 & 2xy \end{vmatrix} = xy^2 = t.$$

Since

$$\frac{\partial(s,t)}{\partial(x,y)} \cdot \frac{\partial(x,y)}{\partial(s,t)} = 1,$$

$$\frac{\partial(x,y)}{\partial(s,t)} = t \qquad \text{so} \qquad \left| \frac{\partial(x,y)}{\partial(s,t)} \right| = \frac{1}{t}$$

So

$$\int_R xy^2\,dA = \int_T t \left| \frac{\partial(x,y)}{\partial(s,t)} \right| ds\,dt = \int_T t\left(\frac{1}{t}\right) ds\,dt = \int_T ds\,dt,$$

where T is the region bounded by $s = 1$, $s = 4$, $t = 1$, $t = 4$.

Then

$$\int_R xy^2\,dA = \int_1^4 ds \int_1^4 dt = 9.$$

Solutions for Chapter 16 Review

Exercises

1. We use Cartesian coordinates, oriented so that the cube is in the first quadrant. See Figure 16.31. Then, if f is an arbitrary function, the integral is

$$\int_0^2 \int_0^3 \int_0^5 f\,dx\,dy\,dz.$$

Other answers are possible. In particular, the order of integration can be changed.

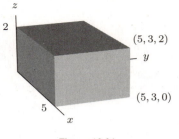

Figure 16.31

5. See Figure 16.32.

9. The region is the half cylinder in Figure 16.33.

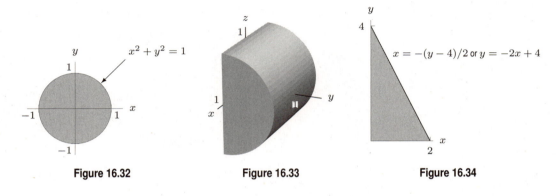

Figure 16.32   **Figure 16.33**   **Figure 16.34**

13. (a) See Figure 16.34.

(b) $\int_0^2 \int_0^{-2x+4} g(x,y)\, dy\, dx$.

17. First use integration by parts, with y as the variable, $u = x^2 y$, $u' = x^2$, $v = \frac{\sin (xy)}{x}$, $v' = \cos (xy)$. Then,

$$\int_3^4 \int_0^1 x^2 y \cos (xy)\, dy\, dx = \int_3^4 \left([xy \sin (xy)]_0^1 - \int_0^1 x \sin (xy)\, dy \right) dx$$

$$= \int_3^4 \left(x \sin x + [\cos (xy)]_0^1 \right) dx$$

$$= \int_3^4 (x \sin x + \cos x - 1)\, dx.$$

Now use integration by parts again, with $u = x$, $u' = 1$, $v = -\cos x$, $v' = \sin x$. Then,

$$\int_3^4 (x \sin x + \cos x - 1)\, dx = [-x \cos x]_3^4 + \int_3^4 \cos x\, dx + \int_3^4 (\cos x - 1)\, dx$$

$$= (-x \cos x + 2 \sin x - x)|_3^4$$

$$= -4 \cos 4 + 2 \sin 4 + 3 \cos 3 - 2 \sin 3 - 1.$$

Thus,

$$\int_3^4 \int_0^1 x^2 y \cos (xy)\, dy\, dx = -4 \cos 4 + 2 \sin 4 + 3 \cos 3 - 2 \sin 3 - 1.$$

21. (a) A vertical plane perpendicular to the x-axis: $x = 2$.

(b) A cylinder: $r = 3$.

(c) A sphere: $\rho = \sqrt{3}$.

(d) A cone: $\phi = \pi/4$.

(e) A horizontal plane: $z = -5$.

(f) A vertical half-plane: $\theta = \pi/4$.

25. From Figure 16.35, we have the following iterated integrals:

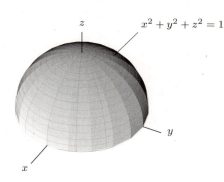

$$x^2 + y^2 + z^2 = 1$$

Figure 16.35

(a) $\displaystyle\int_R f\, dV = \int_{-1}^{1} \int_{-\sqrt{1-x^2}}^{\sqrt{1-x^2}} \int_{0}^{\sqrt{1-x^2-y^2}} f(x,y,z)\, dz\, dy\, dx$

(b) $\displaystyle\int_R f\, dV = \int_{-1}^{1} \int_{-\sqrt{1-y^2}}^{\sqrt{1-y^2}} \int_{0}^{\sqrt{1-x^2-y^2}} f(x,y,z)\, dz\, dx\, dy$

(c) $\displaystyle\int_R f\, dV = \int_{-1}^{1} \int_{0}^{\sqrt{1-y^2}} \int_{-\sqrt{1-y^2-z^2}}^{\sqrt{1-y^2-z^2}} f(x,y,z)\, dx\, dz\, dy$

(d) $\displaystyle\int_R f\, dV = \int_{-1}^{1} \int_{0}^{\sqrt{1-x^2}} \int_{-\sqrt{1-x^2-z^2}}^{\sqrt{1-x^2-z^2}} f(x,y,z)\, dy\, dz\, dx$

(e) $\displaystyle\int_R f\, dV = \int_{0}^{1} \int_{-\sqrt{1-z^2}}^{\sqrt{1-z^2}} \int_{-\sqrt{1-x^2-z^2}}^{\sqrt{1-x^2-z^2}} f(x,y,z)\, dy\, dx\, dz$

(f) $\displaystyle\int_R f\, dV = \int_{0}^{1} \int_{-\sqrt{1-z^2}}^{\sqrt{1-z^2}} \int_{-\sqrt{1-y^2-z^2}}^{\sqrt{1-y^2-z^2}} f(x,y,z)\, dx\, dy\, dz$

Problems

29. The region is a hollow half-sphere, with inner radius $\sqrt{3}$ and outer radius $\sqrt{4} = 2$. See Figure 16.36. In spherical coordinates, the integral is

$$\int_{0}^{\pi} \int_{0}^{\pi} \int_{\sqrt{3}}^{2} \rho^2 \sin\phi\, d\rho\, d\phi\, d\theta.$$

The order of integration can be altered.

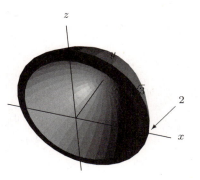

Figure 16.36

33. W is a cylindrical shell, so cylindrical coordinates should be used. See Figure 16.37.

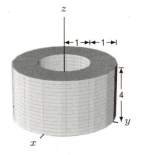

Figure 16.37

$$\int_W \frac{z}{(x^2 + y^2)^{3/2}}\, dV = \int_0^4 \int_0^{2\pi} \int_1^2 \frac{z}{r^3}\, r\, dr\, d\theta\, dz$$

$$= \int_0^4 \int_0^{2\pi} \int_1^2 \frac{z}{r^2}\, dr\, d\theta\, dz$$

$$= \int_0^4 \int_0^{2\pi} \left(-\frac{z}{r}\right)\Big|_1^2\, d\theta\, dz$$

$$= \int_0^4 \int_0^{2\pi} \frac{z}{2}\, d\theta\, dz$$

$$= \int_0^4 \frac{z}{2} \cdot 2\pi\, dz = \frac{1}{2}\pi \cdot z^2 \Big|_0^4 = 8\pi$$

37. Can't tell, since y^3 is both positive and negative for $x < 0$.

41. Zero. You can see this in several ways. One way is to observe that xy is positive on the part of the sphere above and below the first and third quadrants (where x and y are of the same sign) and negative (of equal absolute value) on the part of the sphere above and below the second and fourth quadrants (where x and y have opposite signs). These add up to zero in the integral of xy over all of W.

Another way to see that the integral is zero is to write the triple integral as an iterated integral, say integrating first with respect to x. For fixed y and z, the x-integral is over an interval symmetric about 0. The integral of x over such an interval is zero. If any of the inner integrals in an iterated integral is zero, then the triple integral is zero.

45. Negative. Since $z^2 - 1 \leq 0$ in the sphere, its integral is negative.

49. The depth of the lake is given in meters and the diameter in kilometers. We should work with a single unit of length. In this solution we work with kilometers, but meters would work just as well.

The shape of the lake suggests integration in polar coordinates, with r km measured from the center of the island. Thus $t = r - 1$ is the distance in kilometers from the island when r varies between 1 and 5. The depth of the lake r km from the center of the island is

$$\text{Depth} = \frac{100(r-1)(4-(r-1))}{1000} = -\frac{1}{2} + \frac{3r}{5} - \frac{r^2}{10} \text{ km.}$$

$$\text{Volume of the lake} = \int_0^{2\pi} \int_1^5 \left(-\frac{1}{2} + \frac{3r}{5} - \frac{r^2}{10}\right) r\, dr\, d\theta = \frac{32\pi}{5} = 20.1 \text{ km}^3.$$

53. (a) $\displaystyle\int_0^{2\pi} \int_0^{\pi} \int_1^2 \rho^2 \sin\phi\, d\rho\, d\phi\, d\theta.$

(b) $\displaystyle\int_0^{2\pi} \int_0^2 \int_{-\sqrt{4-r^2}}^{\sqrt{4-r^2}} r\, dz\, dr\, d\theta - \int_0^{2\pi} \int_0^1 \int_{-\sqrt{1-r^2}}^{\sqrt{1-r^2}} r\, dz\, dr\, d\theta.$

57. The region of integration is shown in Figure 16.38, and the mass of the given solid is given by

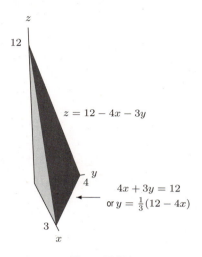

Figure 16.38

$$
\begin{aligned}
\text{Mass} &= \int_R \delta \, dV \\
&= \int_0^3 \int_0^{\frac{1}{3}(12-4x)} \int_0^{12-4x-3y} x^2 \, dz \, dy \, dx \\
&= \int_0^3 \int_0^{\frac{1}{3}(12-4x)} x^2 z \Big|_{z=0}^{z=12-4x-3y} \, dy \, dx \\
&= \int_0^3 \int_0^{\frac{1}{3}(12-4x)} x^2(12-4x-3y) \, dy \, dx \\
&= \int_0^3 x^2 \left(12y - 4xy - \frac{3}{2}y^2\right) \Big|_0^{y=\frac{1}{3}(12-4x)} \, dx \\
&= \left(8x^3 - 4x^4 + \frac{8}{15}x^5\right) \Big|_0^3 \\
&= \frac{108}{5}.
\end{aligned}
$$

61. The plane $(x/p) + (y/q) + (z/r) = 1$ cuts the axes at the points $(p, 0, 0)$; $(0, q, 0)$; $(0, 0, r)$. Since p, q, r are positive, the region between this plane and the coordinate planes is a pyramid in the first octant. Solving for z gives

$$
z = r\left(1 - \frac{x}{p} - \frac{y}{q}\right) = r - \frac{rx}{p} - \frac{ry}{q}.
$$

The volume, V, is given by the double integral

$$
V = \int_R \left(r - \frac{rx}{p} - \frac{ry}{q}\right) dA,
$$

where R is the region shown in Figure 16.39. Thus

$$
V = \int_0^p \int_0^{q-qx/p} \left(r - \frac{rx}{p} - \frac{ry}{q}\right) dy \, dx
$$

$$= \int_0^p \left(\left(ry - \frac{rxy}{p} - \frac{ry^2}{2q} \right) \Big|_{y=0}^{y=q-qx/p} \right) dx$$

$$= \int_0^p \left(r \left(q - \frac{qx}{p} \right) - \frac{r}{p} x \left(q - \frac{qx}{p} \right) - \frac{r}{2q} \left(q - \frac{qx}{p} \right)^2 \right) dx$$

$$= \int_0^p rq - \frac{2rqx}{p} + \frac{rqx^2}{p^2} - \frac{rq^2}{2q} \left(1 - \frac{2x}{p} + \frac{x^2}{p^2} \right) dx$$

$$= \left(rqx - \frac{rqx^2}{p} + \frac{rqx^3}{p^2 3} - \frac{rqx}{2} + \frac{rqx^2}{p2} - \frac{rqx^3}{2p^2 3} \right) \Big|_0^p$$

$$= pqr - pqr + \frac{pqr}{3} - \frac{pqr}{2} + \frac{pqr}{2} - \frac{pqr}{6} = \frac{pqr}{6}.$$

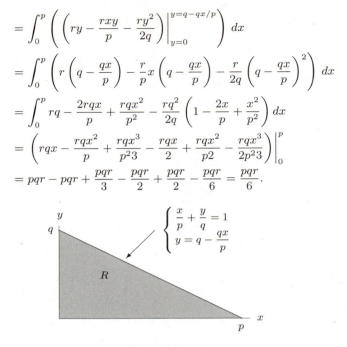

Figure 16.39

65. We must first decide on coordinates. We pick Cartesian coordinates with the smaller sphere centered at the origin, the larger one centered at $(0, 0, -1)$. A vertical cross-section of the region in the xz-plane is shown in Figure 16.40. The smaller sphere has equation $x^2 + y^2 + z^2 = 1$. The larger sphere has equation $x^2 + y^2 + (z + 1)^2 = 2$.

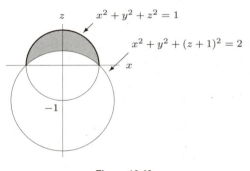

Figure 16.40

Let R represent the region in the xy-plane which lies directly underneath (or above) the region whose volume we want. The curve bounding this region is a circle, and we find its equation by solving the system:

$$x^2 + y^2 + z^2 = 1$$
$$x^2 + y^2 + (z + 1)^2 = 2$$

Subtracting the equations gives

$$(z + 1)^2 - z^2 = 1$$
$$2z + 1 = 1$$
$$z = 0.$$

Since $z = 0$, the two surfaces intersect in the xy-plane in the circle $x^2 + y^2 = 1$. Thus R is $x^2 + y^2 \leq 1$.

The top half of the small sphere is represented by $z = \sqrt{1 - x^2 - y^2}$; the top half of the large sphere is represented by $z = -1 + \sqrt{2 - x^2 - y^2}$. Thus the volume is given by

$$\text{Volume} = \int_{-1}^{1} \int_{-\sqrt{1-x^2}}^{\sqrt{1-x^2}} \int_{-1+\sqrt{2-x^2-y^2}}^{\sqrt{1-x^2-y^2}} dz\,dy\,dx.$$

Starting to evaluate the integral, we get

$$\text{Volume} = \int_{-1}^{1} \int_{-\sqrt{1-x^2}}^{\sqrt{1-x^2}} (\sqrt{1 - x^2 - y^2} + 1 - \sqrt{2 - x^2 - y^2})\,dy\,dx.$$

We simplify the integral by converting to polar coordinates

$$\text{Volume} = \int_{0}^{2\pi} \int_{0}^{1} \left(\sqrt{1 - r^2} + 1 - \sqrt{2 - r^2} \right) r\,dr\,d\theta$$
$$= \int_{0}^{2\pi} \left(-\frac{(1 - r^2)^{3/2}}{3} + \frac{r^2}{2} + \frac{(2 - r^2)^{3/2}}{3} \right) \Bigg|_{0}^{1} d\theta$$
$$= 2\pi \left(\frac{1}{2} + \frac{1}{3} - \left(-\frac{1}{3} + \frac{2^{3/2}}{3} \right) \right) = 2\pi \left(\frac{7}{6} - \frac{2\sqrt{2}}{3} \right) = 1.41.$$

69. The outer circle is a semicircle of radius 4. This is shown in Figure 16.41, with center at D. Thus, $CE = 2$ and $DC = 2$, while $AD = 4$. Notice that angle ADO is a right angle.

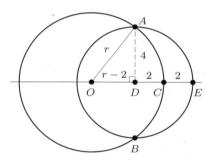

Figure 16.41

Suppose the large circle has center O and radius r. Then $OA = r$ and $OD = OC - DC = r - 2$. Applying Pythagoras' Theorem to triangle OAD gives

$$r^2 = 4^2 + (r - 2)^2$$
$$r^2 = 16 + r^2 - 4r + 4$$
$$r = 5.$$

If we put the origin at O, the equation of the large circle is $x^2 + y^2 = 25$. In the same coordinates, the equation of the small circle, which has center at $D = (3, 0)$, is $(x - 3)^2 + y^2 = 16$. The right hand side of the two circles are given by

$$x = \sqrt{25 - y^2} \quad \text{and} \quad x = 3 + \sqrt{16 - y^2}.$$

Since the y-coordinate of A is 4 and the y-coordinate of B is -4, we have

$$\text{Area} = \int_{-4}^{4} \int_{\sqrt{25-y^2}}^{3+\sqrt{16-y^2}} 1\,dx\,dy$$
$$= \int_{-4}^{4} (3 + \sqrt{16 - y^2} - \sqrt{25 - y^2})\,dy$$
$$= 13.95.$$

CAS Challenge Problems

73. In Cartesian coordinates the integral is

$$\int_D \sqrt[3]{x^2 + y^2}\, dA = \int_{-1}^{1} \int_{-\sqrt{1-x^2}}^{\sqrt{1-x^2}} \sqrt[3]{x^2 + y^2}\, dydx.$$

In polar coordinates it is

$$\int_D \sqrt[3]{x^2 + y^2}\, dA = \int_0^{2\pi} \int_0^1 \sqrt[3]{r^2}\, rdrd\theta = \int_0^{2\pi} \int_0^1 r^{5/3}\, drd\theta$$

$$= \int_0^{2\pi} \frac{3}{8}\, d\theta = \frac{6\pi}{7}$$

The Cartesian coordinate version requires the use of a computer algebra system. Some CASs may be able to handle it and may give the answer in terms of functions called hypergeometric functions. To compare the answers are the same you may need to ask the CAS to give a numerical value for the answer. It's possible your CAS will not be able to handle the integral at all.

CHECK YOUR UNDERSTANDING

1. **False.** For example, if $f(x, y) < 0$ for all (x, y) in the region R, then $\int_R f\, dA$ is negative.

5. **True.** The double integral is the limit of the sum $\displaystyle\sum_{\Delta A \to 0} \rho(x, y)\Delta A$. Each of the terms $\rho(x, y)\Delta A$ is an approximation of the total population inside a small rectangle of area ΔA. Thus the limit of the sum of all of these numbers as $\Delta A \to \infty$ gives the total population of the region R.

9. **False.** There is no reason to expect this to be true, since the behavior of f on one half of R can be completely unrelated to the behavior of f on the other half. As a counterexample, suppose that f is defined so that $f(x, y) = 0$ for points (x, y) lying in S, and $f(x, y) = 1$ for points (x, y) lying in the part of R that is not in S. Then $\int_S f\, dA = 0$, since $f = 0$ on all of S. To evaluate $\int_R f\, dA$, note that $f = 1$ on the square S_1 which is $0 \leq x \leq 1, 1 \leq y \leq 2$. Then $\int_R f\, dA = \int_{S_1} f\, dA = \text{Area}(S_1) = 1$, since $f = 0$ on S.

13. **True.** For any point in the region of integration we have $1 \leq x \leq 2$, and so y is between the positive numbers 1 and 8.

17. **False.** The given limits describe only the upper half disk where $y \geq 0$. The correct limits are $\int_{-a}^{a} \int_{-\sqrt{a^2-x^2}}^{\sqrt{a^2-x^2}} f\, dydx$.

21. **False.** The integral gives the total mass of the material contained in W.

25. **True.** Both sets of limits describe the solid region lying above the rectangle $-1 \leq x \leq 1, 0 \leq y \leq 1, z = 0$ and below the parabolic cylinder $z = 1 - x^2$.

29. **False.** As a counterexample, let W_1 be the solid cube $0 \leq x \leq 1, 0 \leq y \leq 1, 0 \leq z \leq 1$, and let W_2 be the solid cube $-\frac{1}{2} \leq x \leq 0, -\frac{1}{2} \leq y \leq 0, -\frac{1}{2} \leq z \leq 0$. Then volume($W_1$) = 1 and volume($W_2$) = $\frac{1}{8}$. Now if $f(x, y, z) = -1$, then $\int_{W_1} f\, dV = 1 \cdot -1$ which is less than $\int_{W_2} f\, dV = \frac{1}{8} \cdot -1$.

CHAPTER SEVENTEEN

Solutions for Section 17.1

Exercises

1. We want the bottom half of a semicircle of radius 1 centered at $(0, 1)$. The equations $x = \cos t$, $y = 1 + \sin t$ describe clockwise motion in this circle, passing $(-1, 1)$ when $t = \pi$ and $(1, 1)$ when $t = 2\pi$. So a possible parameterization is

$$x = \cos t, \quad y = 1 + \sin t, \quad \pi \leq t \leq 2\pi.$$

5. Since we are moving on the y-axis, $x = 0$, and y goes from -2 to 1. Thus a possible parameterization is

$$x = 0, \quad y = t, \quad -2 \leq t \leq 1.$$

9. One possible parameterization is

$$x = -3 + 2t, \quad y = 4 + 2t, \quad z = -2 - 3t.$$

13. The displacement vector from the first point to the second is $\vec{v} = 4\vec{i} - 5\vec{j} - 3\vec{k}$. The line through point $(1, 5, 2)$ and with direction vector $\vec{v} = 4\vec{i} - 5\vec{j} - 3\vec{k}$ is given by parametric equations

$$x = 1 + 4t,$$
$$y = 5 - 5t,$$
$$z = 2 - 3t.$$

Other parameterizations of the same line are also possible.

17. The line passes through $(3, 0, 0)$ and $(0, 0, -5)$. The displacement vector from the first of these points to the second is $\vec{v} = -3\vec{i} - 5\vec{k}$. The line through point $(3, 0, 0)$ and with direction vector $\vec{v} = -3\vec{i} - 5\vec{k}$ is given by parametric equations

$$x = 3 - 3t,$$
$$y = 0,$$
$$z = -5t.$$

Other parameterizations of the same line are also possible.

21. The circle lies in the plane $z = 2$, so one possible answer is

$$x = 3\cos t, \quad y = 3\sin t, \quad z = 2.$$

25. The xy-plane is $z = 0$, so a possible answer is

$$x = t^2, \quad y = t, \quad z = 0.$$

29. Since its diameters are parallel to the y and z-axes and its center is in the yz-plane, the ellipse must lie in the yz-plane, $x = 0$. The ellipse with the same diameters centered at the origin would have its y-coordinate range between $-5/2$ and $5/2$ and its z-coordinate range between -1 and 1. Thus this ellipse has equation

$$x = 0, \quad y = \frac{5}{2}\cos t, \quad z = \sin t.$$

To move the center to $(0, 1, -2)$, we add 1 to the equation for y and -2 to the equation for z, so one possible answer for our ellipse is

$$x = 0, \quad y = 1 + \frac{5}{2}\cos t, \quad z = -2 + \sin t.$$

33. The displacement vector between the points is $\vec{u} = 3\vec{i} + 5\vec{k}$, so a possible parameterization of the line is

$$x = -1 + 3t, \quad y = 2, \quad z = -3 + 5t.$$

37. The line segment PQ has length 10, so it must be a diameter of the circle. The center of the circle is therefore the midpoint of PQ, which is the point $(5, 0)$. The upper arc of the circle between P and Q can be parameterized as follows:

$$\vec{r}(t) = 5\vec{i} + 5(-\cos t\vec{i} + \sin t\vec{j}), \quad 0 \le t \le \pi.$$

The lower arc can be parameterized as follows:

$$\vec{r}(t) = 5\vec{i} + 5(\cos t\vec{i} + \sin t\vec{j}), \quad \pi \le t \le 2\pi.$$

Problems

41. We find the parameterization in terms of the displacement vector $\overrightarrow{OP} = 2\vec{i} + 5\vec{j}$ from the origin to the point P and the displacement vector $\overrightarrow{PQ} = 10\vec{i} + 4\vec{j}$ from P to Q.
$$\vec{r}(t) = \overrightarrow{OP} + (t - 10)\overrightarrow{PQ} \text{ or } \vec{r}(t) = (2 + (t - 10)10)\vec{i} + (5 + (t - 10)4)\vec{j}$$

45. These equations parameterize a line. Since $(3 + t) + (2t) + 3(1 - t) = 6$, we have $x + y + 3z = 6$. Similarly, $x - y - z = (3 + t) - 2t - (1 - t) = 2$. That is, the curve lies entirely in the plane $x + y + 3z = 6$ and in the plane $x - y - z = 2$. Since the normals to the two planes, $\vec{n_1} = \vec{i} + \vec{j} + 3\vec{k}$ and $\vec{n_2} = \vec{i} - \vec{j} - \vec{k}$ are not parallel, the line is the intersection of two nonparallel planes, which is a straight line in 3-dimensional space.

49. The coefficients of t in the parameterizations show that line $\vec{r}_1$ is parallel to the vector $-3\vec{i} + 2\vec{j} + \vec{k}$ and line $\vec{r}_2$ is parallel to $-6\vec{i} + 4\vec{j} + 3\vec{k}$. Since these vectors are not parallel, the lines are not parallel, so the lines are different.

53. Add the two equations to get $2x + 3z = 5$, or $x = -\frac{3}{2}z + \frac{5}{2}$. Subtract the two equations to get $2y - z = 1$, or $y = \frac{1}{2}z + \frac{1}{2}$. So a possible parameterization is

$$x = -\frac{3}{2}t + \frac{5}{2}, \quad y = \frac{1}{2}t + \frac{1}{2}, \quad z = t.$$

57. (a) Both paths are straight lines, the first passes through the point $(-1, 4, -1)$ in the direction of the vector $\vec{i} - \vec{j} + 2\vec{k}$ and the second passes through $(-7, -6, -1)$ in the direction of the vector $2\vec{i} + 2\vec{j} + \vec{k}$. The two paths are not parallel.
   (b) Is there a time t when the two particles are at the same place at the same time? If so, then their coordinates will be the same, so equating coordinates we get

$$-1 + t = -7 + 2t$$
$$4 - t = -6 + 2t$$
$$-1 + 2t = -1 + t.$$

Since the first equation is solved by $t = 6$, the second by $t = 10/3$, and the third by $t = 0$, no value of t solves all three equations. The two particles never arrive at the same place at the same time, and so they do not collide.
   (c) Are there any times t_1 and t_2 such that the position of the first particle at time t_1 is the same as the position of the second particle at time t_2? If so then

$$-1 + t_1 = -7 + 2t_2$$
$$4 - t_1 = -6 + 2t_2$$
$$-1 + 2t_1 = -1 + t_2.$$

We solve the first two equations and get $t_1 = 2$ and $t_2 = 4$. This is a solution for the third equation as well, so the three equations are satisfied by $t_1 = 2$ and $t_2 = 4$. At time $t = 2$ the first particle is at the point $(1, 2, 3)$, and at time $t = 4$ the second is at the same point. The paths cross at the point $(1, 2, 3)$, and the first particle gets there first.

61. The helices wind around a cylinder of radius α, which explains the significance of α. As t increases from 0 to 2π, the helix winds once around the cylinder, climbing upward a distance of $2\pi\beta$. Thus β controls how stretched out the helix is in the vertical direction. See Figure 17.1 and Figure 17.2.

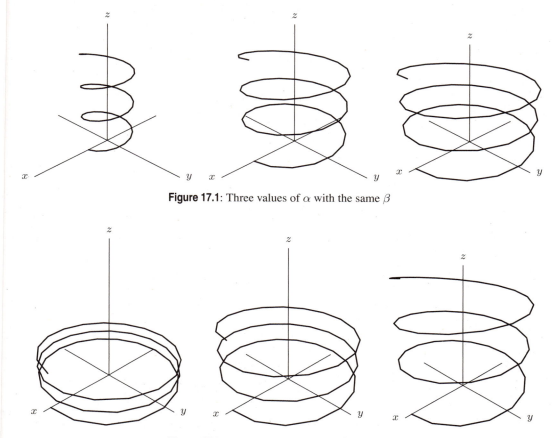

Figure 17.1: Three values of α with the same β

Figure 17.2: Three values of β with the same α

65. The line $\vec{r} = \vec{a} + t\vec{b}$ is parallel to the vector $\vec{b}$ and through the point with position vector $\vec{a}$.

(a) is (vii). The equation $\vec{b} \cdot \vec{r} = 0$ is a plane perpendicular to $\vec{b}$ and satisfied by $(0, 0, 0)$.

(b) is (ii). For any constant k, the equation $\vec{b} \cdot \vec{r} = k$ is a plane perpendicular to $\vec{b}$. If $k = ||\vec{a}|| \neq 0$, the plane does not contain the origin.

(c) is (iv). The equation $(\vec{a} \times \vec{b}) \cdot (\vec{r} - \vec{a}) = 0$ is the equation of a plane which is satisfied by $\vec{r} = \vec{a}$, so the point with position vector $\vec{a}$ lies on the plane. Since $\vec{a} \times \vec{b}$ is perpendicular to $\vec{b}$, the plane is parallel to the line, and therefore it contains the line.

69. (a) If $\vec{n} \cdot \vec{v} = 0$, then $\vec{n}$ and $\vec{v}$ are perpendicular. Since P_1 is perpendicular to $\vec{n}$ and L is parallel to $\vec{v}$, we see that P_1 and L are parallel. In fact, L may lie in the plane.

(b) Since $\vec{n} \times \vec{v}$ is perpendicular to $\vec{n}$ and to $\vec{v}$, the vector $\vec{n} \times \vec{v}$ is parallel to P_1 and perpendicular to L. Thus, P_2, which is perpendicular to $\vec{n} \times \vec{v}$, is

(i) Perpendicular to P_1.

(ii) Parallel to L.

Solutions for Section 17.2

Exercises

1. The velocity vector $\vec{v}$ is given by:

$$\vec{v} = \frac{d(t)}{dt}\vec{i} + \left(\frac{d}{dt}(t^3 - t)\right)\vec{j} = \vec{i} + (3t^2 - 1)\vec{j}.$$

The acceleration vector $\vec{a}$ is given by:

$$\vec{a} = \frac{d\vec{v}}{dt} = \frac{d(1)}{dt}\vec{i} + \left(\frac{d}{dt}(3t^2 - 1)\right)\vec{j} = 6t\vec{j}.$$

5. The velocity vector $\vec{v}$ is given by:

$$\vec{v} = \frac{d}{dt}(3\cos t)\vec{i} + \frac{d}{dt}(4\sin t)\vec{j} = -3\sin t\vec{i} + 4\cos t\vec{j}.$$

The acceleration vector $\vec{a}$ is given by:

$$\vec{a} = \frac{d\vec{v}}{dt} = \frac{d}{dt}(-3\sin t)\vec{i} + \frac{d}{dt}(4\cos t)\vec{j} = -3\cos t\vec{i} - 4\sin t\vec{j}.$$

9. The velocity vector $\vec{v}$ is given by:

$$\vec{v} = \frac{d}{dt}(t)\vec{i} + \frac{d}{dt}(t^2)\vec{j} + \frac{d}{dt}(t^3)\vec{k}$$
$$= \vec{i} + 2t\vec{j} + 3t^2\vec{k}.$$

The speed is given by:

$$\|\vec{v}\| = \sqrt{1 + 4t^2 + 9t^4}.$$

Now $\|\vec{v}\|$ is never zero since $1 + 4t^2 + 9t^4 \geq 1$ for all t. Thus, the particle never stops.

13. We have

$$\text{Length} = \int_1^2 \sqrt{(x'(t))^2 + (y'(t))^2 + (z'(t))^2}\, dt = \int_1^2 \sqrt{5^2 + 4^2 + (-1)^2}\, dt = \sqrt{42}.$$

This is the length of a straight line from the point $(8, 5, 2)$ to $(13, 9, 1)$.

17. The velocity vector $\vec{v}$ is

$$\vec{v} = \frac{dx}{dt}\vec{i} + \frac{dy}{dt}\vec{j} + \frac{dz}{dt}\vec{k} = 0\vec{i} + 2(3)\cos(3t)\vec{j} + 2(3)(-\sin(3t))\vec{k}$$
$$= 6\cos(3t)\vec{j} - 6\sin(3t)\vec{k}.$$

The acceleration vector $\vec{a}$ is

$$\vec{a} = \frac{d^2x}{dt^2}\vec{i} + \frac{d^2y}{dt^2}\vec{j} + \frac{d^2z}{dt^2}\vec{k} = 6(3)(-\sin(3t))\vec{j} - 6(3)\cos(3t)\vec{k}$$
$$= -18\sin(3t)\vec{j} - 18\cos(3t)\vec{k}.$$

To check that $\vec{v}$ and $\vec{a}$ are perpendicular, we check that the dot product is zero:

$$\vec{v} \cdot \vec{a} = (6\cos(3t)\vec{j} - 6\sin(3t)\vec{k}) \cdot (-18\sin(3t)\vec{j} - 18\cos(3t)\vec{k})$$
$$= -108\cos(3t)\sin(3t) + 108\sin(3t)\cos(3t) = 0.$$

The speed is

$$\|\vec{v}\| = \|6\cos(3t)\vec{j} - 6\sin(3t)\vec{k}\| = 6\sqrt{\sin^2(3t) + \cos^2(3t)} = 6,$$

and so is constant. The magnitude of the acceleration is

$$\|\vec{a}\| = \| -18\sin(3t)\vec{j} - 18\cos(3t)\vec{k}\| = 18\sqrt{\sin^2(3t) + \cos^2(3t)} = 18,$$

which is also constant.

Problems

21. Plotting the positions on the xy plane and noting their times gives the graph shown in Figure 17.3.

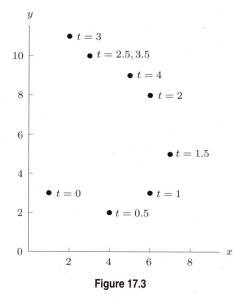

Figure 17.3

(a) We approximate dx/dt by $\Delta x/\Delta t$ calculated between $t = 1.5$ and $t = 2.5$:

$$\frac{dx}{dt} \approx \frac{\Delta x}{\Delta t} = \frac{3 - 7}{2.5 - 1.5} = \frac{-4}{1} = -4.$$

Similarly,

$$\frac{dy}{dt} \approx \frac{\Delta y}{\Delta t} = \frac{10 - 5}{2.5 - 1.5} = \frac{5}{1} = 5.$$

So,

$$\vec{v}(2) \approx -4\vec{i} + 5\vec{j} \quad \text{and} \quad \text{Speed} = \|\vec{v}\| = \sqrt{41}.$$

(b) The particle is moving vertically at about time $t = 1.5$. Note that the particle is momentarily stopped at about $t = 3$; however it is not moving parallel to the y-axis at this instant.

(c) The particle stops at about time $t = 3$ and reverses course.

25. (a) The vector $\overrightarrow{PQ}$ between the points is given by

$$\overrightarrow{PQ} = 2\vec{i} + 5\vec{j} + 3\vec{k}.$$

Since $\|\overrightarrow{PQ}\| = \sqrt{2^2 + 5^2 + 3^2} = \sqrt{38}$, the velocity vector of the motion is

$$\vec{v} = \frac{5}{\sqrt{38}}(2\vec{i} + 5\vec{j} + 3\vec{k}).$$

(b) The motion is along a line starting at the point $(3, 2, -5)$ and with the velocity vector from part (a). The equation of the line is

$$\vec{r} = 3\vec{i} + 2\vec{j} - 5\vec{k} + t\vec{v} = 3\vec{i} + 2\vec{j} - 5\vec{k} + \frac{5}{\sqrt{38}}(2\vec{i} + 5\vec{j} + 3\vec{k})t,$$

so

$$x = 3 + \frac{10}{\sqrt{38}}t, \qquad y = 2 + \frac{25}{\sqrt{38}}t, \qquad z = -5 + \frac{15}{\sqrt{38}}t.$$

29. (a) Since $z = 90$ feet when $t = 0$, the tower is 90 feet high.
    (b) The child reaches the bottom when $z = 0$, so $t = 90/5 = 18$ minutes.
    (c) Her velocity is given by
    $$\vec{v} = \frac{d\vec{r}}{dt} = -(10\sin t)\vec{i} + (10\cos t)\vec{j} - 5\vec{k},$$
    so
    $$\text{Speed} = ||\vec{v}|| = \sqrt{(-10\sin t)^2 + (10\cos t)^2 + (-5)^2} = \sqrt{10^2 + 5^2} = \sqrt{125} \text{ ft/min}.$$
    (d) Her acceleration is given by
    $$\vec{a} = \frac{d\vec{v}}{dt} = -(10\cos t)\vec{i} - (10\sin t)\vec{j} \text{ ft/min}^2.$$

33. (a) For any positive constant k, the parameterization
    $$x = -5\sin(kt) \quad y = 5\cos(kt)$$
    moves counterclockwise on a circle of radius 5 starting at the point $(0, 5)$. We choose k to make the period 8 seconds. If $k \cdot 8 = 2\pi$, then $k = \pi/4$ and the parameterization is
    $$x = -5\sin\left(\frac{\pi t}{4}\right) \quad y = 5\cos\left(\frac{\pi t}{4}\right).$$
    (b) Since it takes 8 seconds for the particle to go around the circle
    $$\text{Speed} = \frac{\text{Circumference of circle}}{8} = \frac{2\pi(5)}{8} = \frac{5\pi}{4} \text{ cm/sec}.$$

37. (a) The center of the wheel moves horizontally, so its y-coordinate will never change; it will equal 1 at all times. In one second, the wheel rotates 1 radian, which corresponds to 1 meter on the rim of a wheel of radius 1 meter, and so the rolling wheel advances at a rate of 1 meter/sec. Thus the x-coordinate of the center, which equals 0 at $t = 0$, will equal t at time t. At time t the center will be at the point $(x, y) = (t, 1)$.
    (b) By time t the spot on the rim will have rotated t radians clockwise, putting it at angle $-t$ as in Figure 17.4. The coordinates of the spot with respect to the center of the wheel are $(\cos(-t), \sin(-t))$. Adding these to the coordinates $(t, 1)$ of the center gives the location of the spot as $(x, y) = (t + \cos t, 1 - \sin t)$. See Figure 17.5.

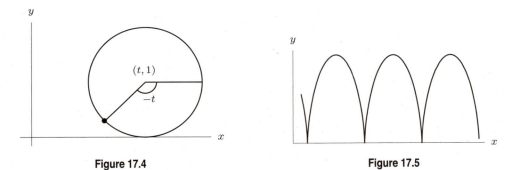

Figure 17.4   **Figure 17.5**

41. (a) Let x represent horizontal displacement (in cm) from some starting point and y the distance (in cm) above the ground. Since
    $$25 \text{ km/hr} = \frac{25 \cdot 10^5}{60^2} = 694.444 \text{ cm/sec},$$
    if t is in seconds, the motion of the center of the pedal is given by
    $$x\vec{i} + y\vec{j} = 694.444t\vec{i} + 30\vec{j}.$$
    The circular motion of your foot relative to the center is described by
    $$h\vec{i} + k\vec{j} = 20\cos(2\pi t)\vec{i} + 20\sin(2\pi t)\vec{j},$$
    so the motion of the light on your foot relative to the ground is described by
    $$x\vec{i} + y\vec{j} = (694.444t + 20\cos(2\pi t))\vec{i} + (30 + 20\sin(2\pi t))\vec{j}.$$

(b) See Figure 17.6.

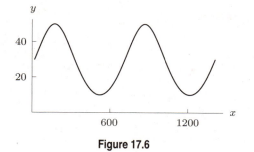

Figure 17.6

(c) Suppose your pedal is rotating with angular velocity ω radians/sec, so that the motion is described by

$$x\vec{i} + y\vec{j} = (694.444t + 20\cos\omega t)\vec{i} + (30 + 20\sin\omega t)\vec{j}.$$

The light moves backward if dx/dt is negative. Since

$$\frac{dx}{dt} = 694.444 - 20\omega\sin\omega t,$$

the minimum value of dx/dt occurs when $\omega t = \pi/2$, and then

$$\frac{dx}{dt} = 694.444 - 20\omega < 0$$

giving

$$\omega \geq 34.722 \text{ radians/sec}.$$

Since there are 2π radians in a complete revolution, an angular velocity of 34.722 radians/sec means $34.722/2\pi \approx 5.526$ revolutions/sec.

45. (a) Using the product rule for differentiation we get

$$\frac{d}{dt}(\vec{r} \cdot \vec{r}) = \vec{r} \cdot \frac{d\vec{r}}{dt} + \frac{d\vec{r}}{dt} \cdot \vec{r} = 2\vec{r} \cdot \frac{d\vec{r}}{dt}.$$

(b) Since $\vec{a}$ is a constant, $d\vec{a}/dt = 0$ so the product rule gives

$$\frac{d}{dt}(\vec{a} \times \vec{r}) = \vec{a} \times \frac{d\vec{r}}{dt}.$$

(c) The product rule gives

$$\frac{d}{dt}(r^3\vec{r}) = r^3\frac{d\vec{r}}{dt} + \frac{d}{dt}(r^3)\vec{r} = r^3\frac{d\vec{r}}{dt} + 3r^2\vec{r}.$$

Solutions for Section 17.3

Exercises

1. (a) Parallel to y-axis.
   (b) Length increasing in x-direction.
   (c) Length not dependent on y.

5. $\vec{V} = x\vec{i}$

9. $\vec{V} = -y\vec{i} + x\vec{j}$

13. See Figure 17.7.

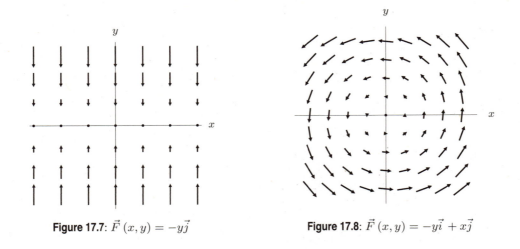

Figure 17.7: $\vec{F}(x, y) = -y\vec{j}$

Figure 17.8: $\vec{F}(x, y) = -y\vec{i} + x\vec{j}$

17. See Figure 17.8.

Problems

21. $\vec{F}(x, y) = a\vec{i} + b\vec{j}$ for any real numbers a and b is a constant vector field. For example, $\vec{F}(x, y) = 3\vec{i} - 4\vec{j}$.

25. If $\vec{F}(x, y) = f(x, y)((1 + y^2)\vec{i} - (x + y)\vec{j})$ where $f(x, y)$ is any function, then $\vec{F} \cdot \vec{G} = 0$, which shows that $\vec{F}$ is perpendicular to $\vec{G}$.

For example $\vec{F}(x, y) = (y + \cos x)((1 + y^2)\vec{i} - (x + y)\vec{j})$.

29. One possible solution is $\vec{F}(x, y) = x\vec{i}$. See Figure 17.9.

Figure 17.9

33. (a) The line l is parallel to the vector $\vec{v} = \vec{i} - 2\vec{j} - 3\vec{k}$. The vector field $\vec{F}$ is parallel to the line when $\vec{F}$ is a multiple of $\vec{v}$. Taking the multiple to be 1 and solving for x, y, z we find a point at which this occurs:

$$x = 1$$
$$x + y = -2$$
$$x - y + z = -3$$

gives $x = 1, y = -3, z = -7$, so a point is $(1, -3, -7)$. Other answers are possible.

(b) The line and vector field are perpendicular if $\vec{F} \cdot \vec{v} = 0$, that is

$$(x\vec{i} + (x+y)\vec{j} + (x-y+z)\vec{k}) \cdot (\vec{i} - 2\vec{j} - 3\vec{k}) = 0$$
$$x - 2x - 2y - 3x + 3y - 3z = 0$$
$$-4x + y - 3z = 0.$$

One point which satisfies this equation is $(0, 0, 0)$. There are many others.

(c) The equation for this set of points is $-4x + y - 3z = 0$. This is a plane through the origin.

37. (a) Dividing a vector $\vec{F}$ by its magnitude always produces the unit vector in the same direction as $\vec{F}$.

(b) Since

$$\|\vec{N}\| = \|(1/F)(-v\vec{i} + u\vec{j})\| = (1/F)\sqrt{v^2 + u^2} = (1/F)F = 1,$$

then $\vec{N}$ is a unit vector. We check that $\vec{N}$ is perpendicular to $\vec{F}$ using the dot product of $\vec{N}$ and $\vec{F}$:

$$\vec{N} \cdot \vec{F} = (1/F)(-v\vec{i} + u\vec{j}) \cdot (u\vec{i} + v\vec{j}) = 0.$$

Which side of $\vec{F}$ does $\vec{N}$ point? The vector $\vec{k}$ is pointing out of the diagram. Since the cross product $\vec{k} \times \vec{F}$ is perpendicular to both $\vec{k}$ and $\vec{F}$, then $\vec{N}$ lies in the xy-plane and points at a right angle to the direction of $\vec{F}$. By the right-hand rule, $\vec{N}$ points to the left as shown in the figure.

Solutions for Section 17.4

Exercises

1. Since $x'(t) = 3$ and $y'(t) = 0$, we have $x = 3t + x_0$ and $y = y_0$. Thus, the solution curves are $y = $ constant.

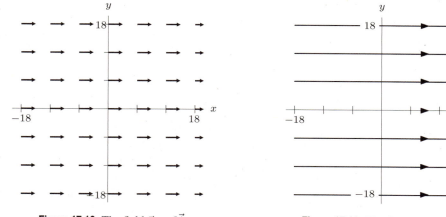

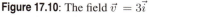

Figure 17.10: The field $\vec{v} = 3\vec{i}$

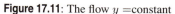

Figure 17.11: The flow $y = $ constant

5. As

$$\vec{v}(t) = \frac{dx}{dt}\vec{i} + \frac{dy}{dt}\vec{j},$$

the system of differential equations is

$$\begin{cases} \frac{dx}{dt} = x \\ \frac{dy}{dt} = 0. \end{cases}$$

Since

$$\frac{d}{dt}(x(t)) = \frac{d}{dt}(ae^t) = x$$

and

$$\frac{d}{dt}(y(t)) = \frac{d}{dt}(b) = 0,$$

the given flow satisfies the system. The solution curves are the horizontal lines $y = b$. See Figures 17.12 and 17.13.

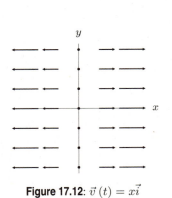

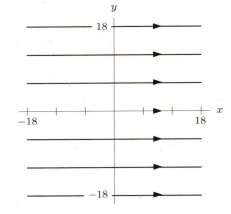

Figure 17.12: $\vec{v}(t) = x\vec{i}$

Figure 17.13: The flow $x(t) = ae^t$, $y(t) = b$

9. As

$$\vec{v}(t) = \frac{dx}{dt}\vec{i} + \frac{dy}{dt}\vec{j},$$

the system of differential equations is

$$\begin{cases} \frac{dx}{dt} = y \\ \frac{dy}{dt} = x. \end{cases}$$

Since

$$\frac{dx(t)}{dt} = \frac{d}{dt}[a(e^t + e^{-t})] = a(e^t - e^{-t}) = y(t)$$

and

$$\frac{dy(t)}{dt} = \frac{d}{dt}[a(e^t - e^{-t})] = a(e^t + e^{-t}) = x(t),$$

the given flow satisfies the system. By eliminating the parameter t in $x(t)$ and $y(t)$, the solution curves obtained are $x^2 - y^2 = 4a^2$. See Figures 17.14 and 17.15.

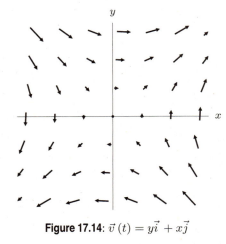

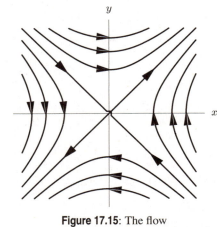

Figure 17.14: $\vec{v}(t) = y\vec{i} + x\vec{j}$

Figure 17.15: The flow
$x(t) = a(e^t + e^{-t}), \ y(t) = a(e^t - e^{-t})$

Problems

13. This corresponds to area C in Figure 17.16.

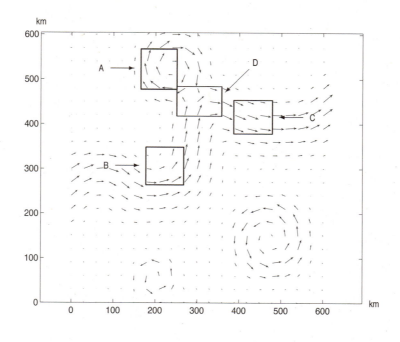

Figure 17.16

17. (a) Perpendicularity is indicated by zero dot product. We have $\vec{v} \cdot \text{grad } H = (-H_y\vec{i} + H_x\vec{j}) \cdot (H_x\vec{i} + H_y\vec{j}) = 0$.
(b) If $\vec{r}(t) = x(t)\vec{i} + y(t)\vec{j}$ is a flow line we have, using the chain rule,

$$\frac{d}{dt}H(x(t), y(t)) = H_x\frac{dx}{dt} + H_y\frac{dy}{dt} = H_x(-H_y) + H_y(H_x) = 0.$$

Thus $H(x(t), y(t))$ is constant which shows that a flow line stays on a single level curve of H.

For a different solution, use geometric reasoning. The vector field $\vec{v}$ is tangent to the level curves of H because, by part (a), $\vec{v}$ and the level curves are both perpendicular to the same vector field grad H. Thus the level curves of H and the flowlines of $\vec{v}$ run in the same direction.

21. Let $\vec{r}(t) = x(t)\vec{i} + y(t)\vec{j}$ be a flow line of $\vec{v}$. If $f(x, y)$ has the same value at all points $(x(t), y(t))$ then the flow line lies on a level curve of f. We can check whether

$$g(t) = f(x(t), y(t)) = x(t)^2 - y(t)^2$$

is constant by computing the derivative $g'(t)$. Since $\vec{v} = y\vec{i} + x\vec{j}$, we have $dx/dt = y$ and $dy/dt = x$. Thus,

$$g'(t) = 2x\frac{dx}{dt} - 2y\frac{dy}{dt} = 2xy - 2yx = 0$$

and g is constant. This means that the flow line lies on a level curve of f. The flow lines are parameterized hyperbolas with equation $x^2 - y^2 = c$.

Solutions for Section 17.5

Exercises

1. A horizontal disk of radius 5 in the plane $z = 7$.

5. Since $z = r = \sqrt{x^2 + y^2}$, we have a cone around the z-axis. Since $0 \le r \le 5$, we have $0 \le z \le 5$, so the cone has height and maximum radius of 5.

9. The top half of the sphere ($z \ge 0$).

Problems

13. Two vectors in the plane containing $P = (0, 0, 0)$, $Q = (1, 2, 3)$, and $R = (2, 1, 0)$ are the displacement vectors
$$\vec{v}_1 = \vec{PQ} = \vec{i} + 2\vec{j} + 3\vec{k}$$
$$\vec{v}_2 = \vec{PR} = 2\vec{i} + \vec{j}.$$
Letting $\vec{r}_0 = 0\vec{i} + 0\vec{j} + 0\vec{k} = \vec{0}$ we have the parameterization
$$\vec{r}(s, t) = \vec{r}_0 + s\vec{v}_1 + t\vec{v}_2$$
$$= (s + 2t)\vec{i} + (2s + t)\vec{j} + 3s\vec{k}.$$

17. To parameterize the plane we need two nonparallel vectors $\vec{v}_1$ and $\vec{v}_2$ that are parallel to the plane. Such vectors are perpendicular to the normal vector to the plane, $\vec{n} = \vec{i} + 2\vec{j} + 3\vec{k}$. We can choose any vectors $\vec{v}_1$ and $\vec{v}_2$ such that $\vec{v}_1 \cdot \vec{n} = \vec{v}_2 \cdot \vec{n} = 0$.

One choice is
$$\vec{v}_1 = 2\vec{i} - j \qquad \vec{v}_2 = 3\vec{i} - \vec{k}.$$
Letting $\vec{r}_0 = 5\vec{i} + \vec{j} + 4\vec{k}$ we have the parameterization
$$\vec{r}(s, t) = \vec{r}_0 + s\vec{v}_1 + t\vec{v}_2$$
$$= (5 + 2s + 3t)\vec{i} + (1 - s)\vec{j} + (4 - t)\vec{k}.$$

21. Since you walk 5 blocks east and 1 block west, you walk 5 blocks in the direction of $\vec{v}_1$, and 1 block in the opposite direction. Thus,
$$s = 5 - 1 = 4,$$
Similarly,
$$t = 4 - 2 = 2.$$
Hence
$$x\vec{i} + y\vec{j} + z\vec{k} = (x_0\vec{i} + y_0\vec{j} + z_0\vec{k}) + 4\vec{v_1} + 2\vec{v_2}$$
$$= (x_0\vec{i} + y_0\vec{j} + z_0\vec{k}) + 4(2\vec{i} - 3\vec{j} + 2\vec{k}) + 2(\vec{i} + 4\vec{j} + 5\vec{k})$$
$$= (x_0 + 10)\vec{i} + (y_0 - 4)\vec{j} + (z_0 + 18)\vec{k}.$$
Thus the coordinates are:
$$x = x_0 + 10, \quad y = y_0 - 4, \quad z = z_0 + 18.$$

25. Set up the coordinates as in Figure 17.17. The surface is the revolution surface obtained by revolving the curve shown in Figure 17.18 about the z axis. From the measurements given, we obtain the equation of the curve in Figure 17.18:
$$x = \cos\left(\frac{\pi}{3}z\right) + 3, \qquad 0 \le z \le 48$$

(a) Rotating this around the z-axis, and taking $z = t$ as the parameter, we get the parametric equations
$$x = \left(\cos\left(\frac{\pi}{3}t\right) + 3\right)\cos\theta$$
$$y = \left(\cos\left(\frac{\pi}{3}t\right) + 3\right)\sin\theta$$
$$z = t \qquad 0 \le \theta \le 2\pi, \ 0 \le t \le 48$$

(b) We know that the points in the curve consists of cross-sections of circles parallel to the xy plane and of radius $\cos((\pi/3)z + 3)$. Thus,
$$\text{Area of cross-section} = \pi\left(\cos\left(\frac{\pi}{3}z + 3\right)\right)^2$$
Integrating over z, we get
$$\text{Volume} = \pi\int_0^{48}\left(\cos\frac{\pi}{3}z + 3\right)^2 dz$$
$$= \pi\int_0^{48}\left(\cos^2\frac{\pi}{3}z + 6\cos\frac{\pi}{3}z + 9\right) dz$$
$$= 456\pi \text{ in}^3.$$

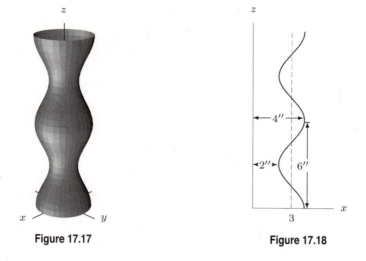

Figure 17.17   **Figure 17.18**

29. Let $(\theta, \pi/2)$ be the original coordinates. If $\theta < \pi$, then the new coordinates will be $(\theta + \pi, \pi/4)$. If $\theta \geq \pi$, then the new coordinates will be $(\theta - \pi, \pi/4)$.

33. The vase obtained by rotating the curve $z = 10\sqrt{x-1}$, $1 \leq x \leq 2$, around the z-axis is shown in Figure 17.19. At height z, the cross-section is a horizontal circle of radius a. Thus, a point on this horizontal circle is given by

$$\vec{r} = a\cos\theta\vec{i} + a\sin\theta\vec{j} + z\vec{k}.$$

However, the radius a varies, so we need to express it in terms of the other parameters θ and z. If you look at the xz-plane, the radius of this circle is given by x, so solving for x in $z = 10\sqrt{x-1}$ gives

$$a = x = \left(\frac{z}{10}\right)^2 + 1.$$

Thus, a parameterization is

$$\vec{r} = \left(\left(\frac{z}{10}\right)^2 + 1\right)\cos\theta\vec{i} + \left(\left(\frac{z}{10}\right)^2 + 1\right)\sin\theta\vec{j} + z\vec{k}$$

so

$$x = \left(\left(\frac{z}{10}\right)^2 + 1\right)\cos\theta, \quad y = \left(\left(\frac{z}{10}\right)^2 + 1\right)\sin\theta, \quad z = z,$$

where $0 \leq \theta \leq 2\pi$, $0 \leq z \leq 10$.

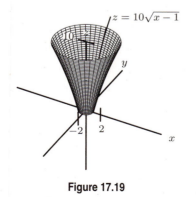

Figure 17.19

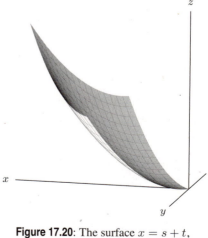

Figure 17.20: The surface $x = s + t$, $y = s - t$, $z = s^2 + t^2$ for $0 \leq s \leq 1$, $0 \leq t \leq 1$

37. (a) From the first two equations we get:

$$s = \frac{x+y}{2}, \qquad t = \frac{x-y}{2}.$$

Hence the equation of our surface is:

$$z = \left(\frac{x+y}{2}\right)^2 + \left(\frac{x-y}{2}\right)^2 = \frac{x^2}{2} + \frac{y^2}{2},$$

which is the equation of a paraboloid.

The conditions: $0 \le s \le 1, 0 \le t \le 1$ are equivalent to: $0 \le x + y \le 2, 0 \le x - y \le 2$. So our surface is defined by:

$$z = \frac{x^2}{2} + \frac{y^2}{2}, \qquad 0 \le x + y \le 2 \quad 0 \le x - y \le 2$$

(b) The surface is shown in Figure 17.20.

Solutions for Chapter 17 Review

Exercises

1. The line has equation

$$\vec{r} = 2\vec{i} - \vec{j} + 3\vec{k} + t(5\vec{i} + 4\vec{j} - \vec{k}),$$

or, equivalently

$$x = 2 + 5t$$
$$y = -1 + 4t$$
$$z = 3 - t.$$

5. The parameterization $x\vec{i} + y\vec{j} = (4 + 4\cos t)\vec{i} + (4 + 4\sin t)\vec{j}$ gives the correct circle, but starts at $(8, 4)$. To start on the x-axis we need

$$x\vec{i} + y\vec{j} = (4 + 4\cos(t - \frac{\pi}{2}))\vec{i} + (4 + 4\sin(t - \frac{\pi}{2}))\vec{j} = (4 + 4\sin t)\vec{i} + (4 - 4\cos t)\vec{j}.$$

9. Since the vector $\vec{n} = \text{grad}(2x - 3y + 5z) = 2\vec{i} - 3\vec{j} + 5\vec{k}$ is perpendicular to the plane, this vector is parallel to the line. Thus the equation of the line is

$$x = 1 + 2t, \quad y = 1 - 3t, \quad z = 1 + 5t.$$

13. See Figure 17.21. The parameterization is

$$\vec{r} = 10\cos\left(\frac{2\pi t}{30}\right)\vec{i} - 10\sin\left(\frac{2\pi t}{30}\right)\vec{j} + 7\vec{k}.$$

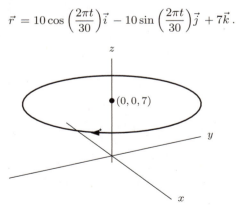

Figure 17.21

17. The velocity vector $\vec{v}$ is given by:

$$\vec{v} = \frac{d}{dt}(2 + 3t^2)\vec{i} + \frac{d}{dt}(4 + t^2)\vec{j} + \frac{d}{dt}(1 - t^2)\vec{k} = 6t\vec{i} + 2t\vec{j} - 2t\vec{k}.$$

21. Vector. Differentiating using the chain rule gives

$$\text{Velocity} = \left(\frac{-\cos t}{2\sqrt{3 + \sin t}}\right)\vec{i} + \left(\frac{-\sin t}{2\sqrt{3 + \cos t}}\right)\vec{j}.$$

25. The direction vectors of the lines, $-\vec{i} + 4\vec{j} - 2\vec{k}$ and $2\vec{i} - 8\vec{j} + 4\vec{k}$, are multiplies of each other (the second is -2 times the first). Thus the lines are parallel. To see if they are the same line, we take the point corresponding to $t = 0$ on the first line, which has position vector $3\vec{i} + 3\vec{j} - \vec{k}$, and see if it is on the second line. So we solve

$$(1 + 2t)\vec{i} + (11 - 8t)\vec{j} + (4t - 5)\vec{k} = 3\vec{i} + 3\vec{j} - \vec{k}.$$

This has solution $t = 1$, so the two lines have a point in common and must be the same line, parameterized in two different ways.

29. The vector field points in a clockwise direction around the origin. Since

$$\left\| \left(\frac{y}{\sqrt{x^2 + y^2}}\right)\vec{i} - \left(\frac{x}{\sqrt{x^2 + y^2}}\right)\vec{j} \right\| = \frac{\sqrt{x^2 + y^2}}{\sqrt{x^2 + y^2}} = 1,$$

the length of the vectors is constant everywhere.

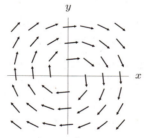

Figure 17.22

Problems

33. (a) A vector field associates a vector to every point in a region of the space. In other words, a vector field is a vector-valued function of position given by $\vec{v} = \vec{f}(\vec{r}) = \vec{f}(x, y, z)$

(b) (i) Yes, $\vec{r} + \vec{a} = (x + a_1)\vec{i} + (y + a_2)\vec{j} + (z + a_3)\vec{k}$ is a vector-valued function of position.

(ii) No, $\vec{r} \cdot \vec{a}$ is a scalar.

(iii) Yes.

(iv) $x^2 + y^2 + z^2$ is a scalar.

37. The displacement vector from $(1, 1, 1)$ to $(2, -1, 3)$ is $\vec{d} = (2\vec{i} - \vec{j} + 3\vec{k}) - (\vec{i} + \vec{j} + \vec{k}) = \vec{i} - 2\vec{j} + 2\vec{k}$ meters. The velocity vector has the same direction as $\vec{d}$ and is given by

$$\vec{v} = \frac{\vec{d}}{5} = 0.2\vec{i} - 0.4\vec{j} + 0.4\vec{k} \text{ meters/sec.}$$

Since $\vec{v}$ is constant, the acceleration $\vec{a} = \vec{0}$.

41. (a) In order for the particle to stop, its velocity $\vec{v} = (dx/dt)\vec{i} + (dy/dt)\vec{j}$ must be zero, so we solve for t such that $dx/dt = 0$ and $dy/dt = 0$, that is

$$\frac{dx}{dt} = 3t^2 - 3 = 3(t-1)(t+1) = 0,$$

$$\frac{dy}{dt} = 2t - 2 = 2(t-1) = 0.$$

The value $t = 1$ is the only solution. Therefore, the particle stops when $t = 1$ at the point $(t^3 - 3t, \ t^2 - 2t)|_{t=1} = (-2, -1)$.

(b) In order for the particle to be traveling straight up or down, the x-component of the velocity vector must be 0. Thus, we solve $dx/dt = 3t^2 - 3 = 0$ and obtain $t = \pm 1$. However, at $t = 1$ the particle has no vertical motion, as we saw in part (a). Thus, the particle is moving straight up or down only when $t = -1$. Since the velocity at time $t = -1$ is

$$\vec{v}(-1) = \frac{dx}{dt}\bigg|_{t=-1} \vec{i} + \frac{dy}{dt}\bigg|_{t=-1} \vec{j} = -4\vec{j},$$

the motion is straight down. The position at that time is $(t^3 - 3t, \ t^2 - 2t)|_{t=-1} = (2, 3)$.

(c) For horizontal motion we need $dy/dt = 0$. That happens when $dy/dt = 2t - 2 = 0$, and so $t = 1$. But from part (a) we also have $dx/dt = 0$ also at $t = 1$, so the particle is not moving at all when $t = 1$. Thus, there is no time when the motion is horizontal.

45. (a) $f_x = \dfrac{\left[2x(x^2 + y^2) - 2x(x^2 - y^2)\right]}{(x^2 + y^2)^2} = \dfrac{4xy^2}{(x^2 + y^2)^2}.$

$f_y = \dfrac{\left[-2y(x^2 + y^2) - 2y(x^2 - y^2)\right]}{(x^2 + y^2)^2} = \dfrac{-4yx^2}{(x^2 + y^2)^2}.$

$\nabla f(1,1) = \vec{i} - \vec{j}$, i.e., south-east.

(b) We need a vector $\vec{u}$ such that $\nabla f(1,1) \cdot \vec{u} = 0$, i.e., such that $(\vec{i} - \vec{j}) \cdot \vec{u} = 0$. The vector $\vec{u} = \vec{i} + \vec{j}$ clearly works; so does $\vec{u} = -\vec{i} - \vec{j}$. Dividing by the length to get a unit vector, we have $\vec{u} = \frac{1}{\sqrt{2}}\vec{i} + \frac{1}{\sqrt{2}}\vec{j}$ or $\vec{u} = -\frac{1}{\sqrt{2}}\vec{i} - \frac{1}{\sqrt{2}}\vec{j}$.

(c) f is a function of x and y, which in turn are functions of t. Thus, the chain rule can be used to show how f changed with t.

$$\frac{df}{dt} = \frac{\partial f}{\partial x} \cdot \frac{dx}{dt} + \frac{\partial f}{\partial y} \cdot \frac{dy}{dt} = \frac{4xy^2}{(x^2 + y^2)^2} \cdot 2e^{2t} - \frac{4x^2 y}{(x^2 + y^2)^2} \cdot (6t^2 + 6).$$

At $t = 0$, $x = 1, y = 1$; so, $\dfrac{df}{dt} = \dfrac{4}{4} \cdot 2 - \dfrac{4}{4} \cdot 6 = -4$.

49. Note that uniform circular motion is possible with the given conditions, since $\vec{v} \cdot \vec{a} = 0$ shows that the velocity and acceleration vectors are perpendicular.

For uniform circular motion in an orbit of radius R, we have $\|\vec{a}\| = \|\vec{v}\|^2/R$. Thus $R = \|\vec{v}\|^2/\|\vec{a}\| = 52/\sqrt{13}$ for both parts (a) and (b).

The center of the orbit is at distance R in the direction of the acceleration vector from the point P on the orbit. The vector

$$R\frac{\vec{a}}{\|\vec{a}\|} = \frac{52}{\sqrt{13}} \frac{2\vec{i} + 3\vec{j}}{\sqrt{13}} = 8\vec{i} + 12\vec{j}$$

thus extends from the point P to the center of the orbit.

(a) The center of the orbit is at the point $(0 + 8, 0 + 12) = (8, 12)$

(b) The center of the orbit is at the point $(10 + 8, 50 + 12) = (18, 62)$

53. At time t the particle has polar coordinates $r = \|\vec{r}(t)\| = at$ and $\theta = \omega t$. At time t, the ray from the origin to the particle is at angle ωt radians from the positive x-axis. The ray is therefore rotating at a rate of ω radians per unit time. The parameter ω is the rate of change of the polar angle θ of the particle measured in radians per unit time. The larger ω is, the quicker the particle completes a complete revolution (a 360° trip) around the origin. At time t, the particle is at distance at from the origin. Thus a equals the rate of change of the particle's distance from the origin. The larger a is, the faster the particle moves away from the origin.

57. (a) The cone of height h, maximum radius a, vertex at the origin and opening upward is shown in Figure 17.23.

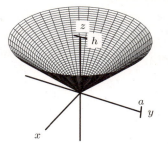

Figure 17.23

By similar triangles, we have

$$\frac{r}{z} = \frac{a}{h},$$

so

$$z = \frac{hr}{a}.$$

Therefore, one parameterization is

$$x = r\cos\theta, \qquad 0 \le r \le a,$$
$$y = r\sin\theta, \qquad 0 \le \theta < 2\pi,$$
$$z = \frac{hr}{a}.$$

(b) Since $r = az/h$, we can write the parameterization in part (a) as

$$x = \frac{az}{h}\cos\theta, \qquad 0 \le z \le h,$$
$$y = \frac{az}{h}\sin\theta, \qquad 0 \le \theta < 2\pi,$$
$$z = z.$$

61. The parameterization for a sphere of radius a using spherical coordinates is

$$x = a\sin\phi\cos\theta, \quad y = a\sin\phi\sin\theta, \quad z = a\cos\phi.$$

Think of the ellipsoid as a sphere whose radius is different along each axis and you get the parameterization:

$$\begin{cases} x = a\sin\phi\cos\theta, & 0 \le \phi \le \pi, \\ y = b\sin\phi\sin\theta, & 0 \le \theta \le 2\pi, \\ z = c\cos\phi. \end{cases}$$

To check this parameterization, substitute into the equation for the ellipsoid:

$$\frac{x^2}{a^2} + \frac{y^2}{b^2} + \frac{z^2}{c^2} = \frac{a^2\sin^2\phi\cos^2\theta}{a^2} + \frac{b^2\sin^2\phi\sin^2\theta}{b^2} + \frac{c^2\cos^2\phi}{c^2}$$
$$= \sin^2\phi(\cos^2\theta + \sin^2\theta) + \cos^2\phi = 1.$$

CAS Challenge Problems

65. (a)

$$\vec{r} \cdot \vec{F} = (x\vec{i} + y\vec{j}) \cdot (-y(1-y^2)\vec{i} + x(1-y^2)\vec{j})$$
$$= -xy(1-y^2) + yx(1-y^2) = 0$$

This means that the tangent line to the flow line at a point is always perpendicular to the vector from the origin to that point. Hence the flow lines are circles centered at the origin.

(b) The circle $\vec{r}(t) = \cos t\vec{i} + \sin t\vec{j}$ has velocity vector $\vec{v}(t) = -\sin t\vec{i} + \cos t\vec{j} = -y\vec{i} + x\vec{j} = (1-y^2)\vec{F}$. Thus the velocity vector is a scalar multiple of $\vec{F}$, and hence parallel to $\vec{F}$. However, since $\vec{v}(t)$ is not equal to $\vec{F}(\vec{r}(t))$, it is not a flow line.

(c) Using a CAS, we find

$$\vec{v}(t) = -\frac{t}{(1+t^2)^{3/2}}\vec{i} + \left(-\frac{t^2}{(1+t^2)^{3/2}} + \frac{1}{\sqrt{1+t^2}}\right)\vec{j} = -\frac{t}{(1+t^2)^{3/2}}\vec{i} + \frac{1}{(1+t^2)^{3/2}}\vec{j}$$

and

$$\vec{F}(\vec{r}(t)) = -\left(\frac{t\left(1 - \frac{t^2}{1+t^2}\right)}{\sqrt{1+t^2}}\right)\vec{i} + \frac{1 - \frac{t^2}{1+t^2}}{\sqrt{1+t^2}}\vec{j} = -\frac{t}{(1+t^2)^{3/2}}\vec{i} + \frac{1}{(1+t^2)^{3/2}}\vec{j} = \vec{v}(t).$$

Although the circle parameterized in part (b) has velocity vectors parallel to $\vec{F}$ at each point of the circle, its speed is not equal to the magnitude of the vector field. The circle in part (c) is parameterized at the correct speed to be the flow line.

CHECK YOUR UNDERSTANDING

1. False. The y coordinate is zero when $t = 0$, but when $t = 0$ we have $x = 2$ so the curve never passes through $(0, 0)$.

5. False. When $t = 0$, we have $(x, y) = (0, -1)$. When $t = \pi/2$, we have $(x, y) = (-1, 0)$. Thus the circle is being traced out clockwise.

9. True. To find an intersection point, we look for values of s and t that make the coordinates in the first line the same as the coordinates in the second. Setting $x = t$ and $x = 2s$ equal, we see that $t = 2s$. Setting $y = 2 + t$ equal to $y = 1 - s$, we see that $t = -1 - s$. Solving both $t = 2s$ and $t = -1 - s$ yields $t = -\frac{2}{3}, s = -\frac{1}{3}$. These values of s and t will give equal x and y coordinates on both lines. We need to check if the z coordinates are equal also. In the first line, setting $t = -\frac{2}{3}$ gives $z = \frac{7}{3}$. In the second line, setting $s = -\frac{1}{3}$ gives $z = -\frac{1}{3}$. As these are not the same, the lines do not intersect.

13. False. The velocity vector is $\vec{v}(t) = \vec{r}\,'(t) = 2t\vec{i} - \vec{j}$. Then $\vec{v}(-1) = -2\vec{i} - \vec{j}$ and $\vec{v}(1) = 2\vec{i} - \vec{j}$, which are not equal.

17. False. As a counterexample, consider the curve $\vec{r}(t) = t^2\vec{i} + t^2\vec{j}$ for $0 \le t \le 1$. In this case, when t is replaced by $-t$, the parameterization is the same, and is not reversed.

21. True, since the vectors $x\vec{j}$ are parallel to the y-axis.

25. True. Any flow line which stays in the first quadrant has $x, y \to \infty$.

29. True. If (x, y) were a point where the y-coordinate along a flow line reached a relative maximum, then the tangent vector to the flow line, namely $\vec{F}(x, y)$, there would have to be horizontal (or $\vec{0}$), that is its $\vec{j}$ component would have to be 0. But the $\vec{j}$ component of $\vec{F}$ is always 2.

33. False. There is only one parameter, s. The equations parameterize a line.

37. True. If the surface is parameterized by $\vec{r}(s, t)$ and the point has parameters (s_0, t_0) then the parameter curves $\vec{r}(s_0, t)$ and $\vec{r}(s, t_0)$.

41. False. Suppose $\vec{r}(t) = t\vec{i} + t\vec{j}$. Then $\vec{r}\,'(t) = \vec{i} + \vec{j}$ and

$$\vec{r}\,'(t) \cdot \vec{r}(t) = (t\vec{i} + t\vec{j}) \cdot (\vec{i} + \vec{j}) = 2t.$$

So $\vec{r}\,'(t) \cdot \vec{r}(t) \ne 0$ for $t \ne 0$.

CHAPTER EIGHTEEN

Solutions for Section 18.1

Exercises

1. Positive, because the vectors are longer on the portion of the path that goes in the same direction as the vector field.

5. Negative, because the vector field points in the opposite direction to the path.

9. Since $\vec{F}$ is a constant vector field and the curve is a line, $\int_C \vec{F} \cdot d\vec{r} = \vec{F} \cdot \Delta \vec{r}$, where $\Delta \vec{r} = 7\vec{j}$. Therefore,

$$\int_C \vec{F} \cdot d\vec{r} = (3\vec{i} + 4\vec{j}) \cdot 7\vec{j} = 28$$

13. At every point, the vector field is parallel to segments $\Delta \vec{r} = \Delta x \vec{i}$ of the curve. Thus,

$$\int_C \vec{F} \cdot d\vec{r} = \int_2^6 x\vec{i} \cdot dx\vec{i} = \int_2^6 x\,dx = \frac{x^2}{2}\Big|_2^6 = 16.$$

17. Since the curve is along the y-axis, only the $\vec{j}$ component of the vector field contributes to the integral:

$$\int_C (2\vec{j} + 3\vec{k}) \cdot d\vec{r} = \int_C 2\vec{j} \cdot d\vec{r} = 2 \cdot \text{ Length of } C = 2 \cdot 10 = 20.$$

Problems

21. Since it appears that C_1 is everywhere perpendicular to the vector field, all of the dot products in the line integral are zero, hence $\int_{C_1} \vec{F} \cdot d\vec{r} \approx 0$. Along the path C_2 the dot products of $\vec{F}$ with $\Delta \vec{r}_i$ are all positive, so their sum is positive and we have $\int_{C_1} \vec{F} \cdot d\vec{r} < \int_{C_2} \vec{F} \cdot d\vec{r}$. For C_3 the vectors $\Delta \vec{r}_i$ are in the opposite direction to the vectors of $\vec{F}$, so the dot products $\vec{F} \cdot \Delta \vec{r}_i$ are all negative; so, $\int_{C_3} \vec{F} \cdot d\vec{r} < 0$. Thus, we have

$$\int_{C_3} \vec{F} \cdot d\vec{r} < \int_{C_1} \vec{F} \cdot d\vec{r} < \int_{C_2} \vec{F} \cdot d\vec{r}$$

25. The line integral along C_1 is negative, the line integral along C_2 is negative, and the line integral along C_3 appears to be zero.

29. The vector field $\vec{F}$ is in the same direction as C if $b > 0$, so we want $b < 0$. No restriction is needed on c.

33. This vector field is illustrated in Figure 18.1. It is perpendicular to C_2 and C_4 at every point, since $\vec{F}(x, y) \cdot \vec{r}(x, y) = 0$ and C_2 and C_4 are radial line segments, then

$$\int_{C_2} \vec{F} \cdot d\vec{r} = \int_{C_4} \vec{F} \cdot d\vec{r} = 0.$$

Since C_3 is longer than C_1, and the vector field is larger in magnitude along C_3, the line integral along C_3 has greater absolute value than that along C_1. The line integral along C_3 is positive and the line integral along C_1 is negative, so

$$\int_C \vec{F} \cdot d\vec{r} = \int_{C_3} \vec{F} \cdot d\vec{r} + \int_{C_1} \vec{F} \cdot d\vec{r} > 0.$$

See Figure 18.1.

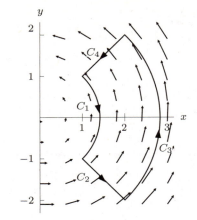

Figure 18.1

37. (a) See Figure 18.2.

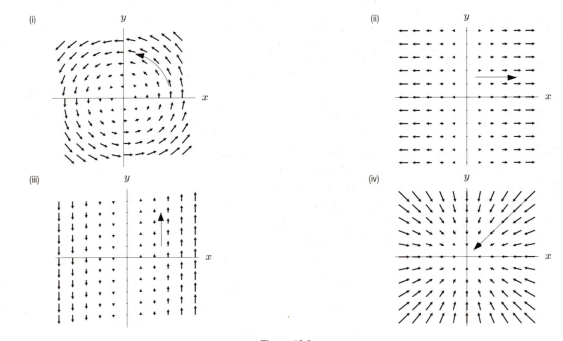

Figure 18.2

(b) For (i) and (iii) a closed curve can be drawn; not for the others.

41. See Figure 18.3. The example chosen is the vector field $\vec{F}(x, y) = y\vec{j}$ and the path C is the line from $(0, -1)$ to $(0, 1)$. Since the vectors are symmetric about the x-axis, the dot products $\vec{F} \cdot \Delta \vec{r}$ cancel out along C to give 0 for the line integral. Many other answers are possible.

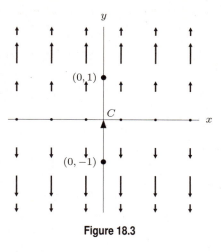

Figure 18.3

45. Let $r = \|\vec{r}\|$. Since $\Delta \vec{r}$ points outward, in the opposite direction to $\vec{F}$, we expect the answer to be negative.

$$\int_C \vec{F} \cdot d\vec{r} = \int_C -\frac{GMm\vec{r}}{r^3} \cdot d\vec{r} = \int_{8000}^{10000} -\frac{GMm}{r^2}\,dr$$

$$= \left.\frac{GMm}{r}\right|_{8000}^{10000} = GMm\left(\frac{1}{10000} - \frac{1}{8000}\right)$$

$$= -2.5 \cdot 10^{-5} GMm.$$

49. In Problem 48 we saw that the surface where the potential is zero is a sphere of radius a. Let S be any sphere centered at the origin, and let P_1 be a point on S, and C_1 a path from P_0 to P_1. If P is any point on S, then P can be reached from P_0 by a path, C, consisting of C_1 followed by C_2, where C_2 is a path from P_1 to P lying entirely on the sphere, S. Then $\int_{C_2} \vec{E} \cdot d\vec{r} = 0$, since $\vec{E}$ is perpendicular to the sphere. So

$$\phi(P) = -\int_C \vec{E} \cdot d\vec{r} = -\int_{C_1} \vec{E} \cdot d\vec{r} - \int_{C_2} \vec{E} \cdot d\vec{r} = -\int_{C_1} \vec{E} \cdot d\vec{r} = \phi(P_1).$$

Thus, ϕ is constant on S. The equipotential surfaces are spheres centered at the origin.

Solutions for Section 18.2

Exercises

1. Only the $\vec{i}$-component contributes to the line integral, so $d\vec{r} = \vec{i}\,dx$ and

$$\int_C (2x\vec{i} + 3y\vec{j}) \cdot d\vec{r} = \int_{(1,0,0)}^{(5,0,0)} (2x\vec{i} + 3y\vec{j}) \cdot \vec{i}\,dx = \int_1^5 2x\,dx = \left.x^2\right|_1^5 = 24.$$

5. Since $\vec{F} = (x^2 + y)\vec{i} + y^3\vec{j}$, the line integral along the third segment, which is parallel to the z-axis, is zero. On the first segment, which is parallel to the y-axis, only the $\vec{j}$-component contributes. On the second segment, which is parallel to the x-axis, only the $\vec{i}$-component contributes. On the first segment $x = 4$ and y varies from 0 to 3; on the second segment $y = 3$ and x varies from 4 to 0. Thus, we have

$$\int_C \vec{F} \cdot d\vec{r} = \int_0^3 ((4^2 + y)\vec{i} + y^3\vec{j}) \cdot \vec{j}\,dy + \int_4^0 ((x^2 + 3)\vec{i} + 3^3\vec{j}) \cdot \vec{i}\,dx$$

$$= \int_0^3 y^3\,dy + \int_4^0 (x^2 + 3)\,dx = \left.\frac{y^4}{4}\right|_0^3 - \left(\frac{x^3}{3} + 3x\right)\Big|_0^4 = \frac{81}{4} - \frac{64}{3} - 12 = -\frac{157}{12}.$$

9. The line can be parameterized by $(1 + 2t, 2 + 2t)$, for $0 \leq t \leq 1$, so the integral looks like

$$
\begin{aligned}
\int_C \vec{F} \cdot d\vec{r} &= \int_0^1 \vec{F}(1 + 2t, 2 + 2t) \cdot (2\vec{i} + 2\vec{j}) \, dt \\
&= \int_0^1 [(1 + 2t)^2 \vec{i} + (2 + 2t)^2 \vec{j}] \cdot (2\vec{i} + 2\vec{j}) \, dt \\
&= \int_0^1 2(1 + 4t + 4t^2) + 2(4 + 8t + 4t^2) \, dt \\
&= \int_0^1 (10 + 24t + 16t^2) \, dt \\
&= (10t + 12t^2 + 16t^3/3) \Big|_0^1 \\
&= 10 + 12 + 16/3 - (0 + 0 + 0) = 82/3
\end{aligned}
$$

13. The curve C is parameterized by

$$
\vec{r} = \cos t \vec{i} + \sin t \vec{j}, \qquad \text{for } 0 \leq t \leq 2\pi,
$$

so,

$$
\vec{r}'(t) = -\sin t \vec{i} + \cos t \vec{j}.
$$

Thus,

$$
\begin{aligned}
\int_C \vec{F} \cdot d\vec{r} &= \int_0^{2\pi} (2\sin t \vec{i} - \sin(\sin t)\vec{j}) \cdot (-\sin t \vec{i} + \cos t \vec{j}) \, dt \\
&= \int_0^{2\pi} (-2\sin^2 t - \sin(\sin t)\cos t) \, dt \\
&= \sin t \cos t - t + \cos(\sin t) \Big|_0^{2\pi} \\
&= -2\pi.
\end{aligned}
$$

17. Since $\vec{r} = x(t)\vec{i} + y(t)\vec{j} + z(t)\vec{k} = t\vec{i} + t^2\vec{j} + t^3\vec{k}$, for $1 \leq t \leq 2$,
we have $\vec{r}'(t) = x'(t)\vec{i} + y'(t)\vec{j} + z'(t)\vec{k} = \vec{i} + 2t\vec{j} + 3t^2\vec{k}$. Then

$$
\begin{aligned}
\int_C \vec{F} \cdot d\vec{r} &= \int_1^2 (t\vec{i} + 2t^3 t^2 \vec{j} + t\vec{k}) \cdot (\vec{i} + 2t\vec{j} + 3t^2\vec{k}) \, dt \\
&= \int_1^2 (t + 4t^6 + 3t^3) \, dt \\
&= \frac{t^2}{2} + \frac{4t^7}{7} + \frac{3t^4}{4} \Big|_1^2 = \frac{2389}{28} \approx 85.32
\end{aligned}
$$

21. $\int_C 3x\,dx - y\sin x\,dy$

25. From $x = t^2$ and $y = t^3$ we get $dx = 2t\,dt$ and $dy = 3t^2\,dt$. Hence

$$
\int_C y\,dx + x\,dy = \int_1^5 t^3(2t)\,dt + t^2(3t^2)\,dt = \int_1^5 5t^4\,dt = 5^5 - 1 = 3124.
$$

Problems

29. (a) Figure 18.4 shows the curves.

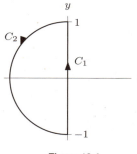

Figure 18.4

(b) On C_1, only the $\vec{j}$ -component of $\vec{F}$ contributes to the integral. There $d\vec{r} = \vec{j}\ dy$, so

$$\int_{C_1} \vec{F} \cdot d\vec{r} = \int_{-1}^{1} y\vec{j} \cdot \vec{j}\ dy = \int_{-1}^{1} y\ dy = \left.\frac{y^2}{2}\right|_{-1}^{1} = 0.$$

On C_2, we have $\vec{r}\,'(t) = -\sin t\vec{i} + \cos t\vec{j}$, so

$$\int_{C_2} \vec{F} \cdot d\vec{r} = \int_{\pi/2}^{3\pi/2} ((\cos t + 3\sin t)\vec{i} + \sin t\vec{j}) \cdot (-\sin t\vec{i} + \cos t\vec{j})\ dt$$

$$= \int_{\pi/2}^{3\pi/2} -\cos t\sin t - 3\sin^2 t + \cos t\sin t\ dt = \int_{\pi/2}^{3\pi/2} -3\sin^2 t\ dt$$

$$= -3\left(\frac{t}{2} - \frac{\sin t\cos t}{2}\right)\Bigg|_{\pi/2}^{3\pi/2} = -\frac{3\pi}{2}.$$

33. (a) The line integral $\int_C (xy\vec{i} + x\vec{j}) \cdot d\vec{r}$ is positive. This follows from the fact that all of the vectors of $xy\vec{i} + x\vec{j}$ at points along C point approximately in the same direction as C (meaning the angles between the vectors and the direction of C are less than $\pi/2$).

(b) Using the parameterization $x(t) = t$, $y(t) = 3t$, with $x'(t) = 1$, $y'(t) = 3$, we have

$$\int_C \vec{F} \cdot d\vec{r} = \int_0^4 \vec{F}(t, 3t) \cdot (\vec{i} + 3\vec{j})\ dt$$

$$= \int_0^4 (3t^2\vec{i} + t\vec{j}) \cdot (\vec{i} + 3\vec{j})\ dt$$

$$= \int_0^4 (3t^2 + 3t)\ dt$$

$$= \left(t^3 + \frac{3}{2}t^2\right)\Bigg|_0^4$$

$$= 88.$$

(c) Figure 18.5 shows the oriented path C', with the "turn around" points P and Q. The particle first travels from the origin to the point P (call this path C_1), then backs up from P to Q (call this path C_2), then goes from Q to the point $(4, 12)$ in the original direction (call this path C_3). See Figure 18.6. Thus, $C' = C_1 + C_2 + C_3$. Along the parts of C_1 and C_2 that overlap, the line integrals cancel, so we are left with the line integral over the part of C_1 that does not overlap with C_2, followed by the line integral over C_3. Thus, the line integral over C' is the same as the line integral over the direct route from the point $(0, 0)$ to the point $(4, 12)$.

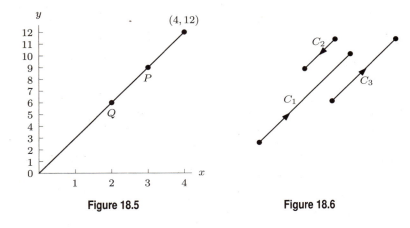

Figure 18.5   **Figure 18.6**

(d) The parameterization

$$(x(t), y(t)) = \left(\frac{1}{3}(t^3 - 6t^2 + 11t), (t^3 - 6t^2 + 11t) \right)$$

has $(x(0), y(0)) = (0, 0)$ and $(x(4), y(4)) = (4, 12)$. The form of the parameterization we were given shows that the second coordinate is always three times the first. Thus all points on the parameterized curve lie on the line $y = 3x$.

We have to do a bit more work to guarantee that all points on the curve lie on the line *between* the point $(0, 0)$ and the point $(4, 12)$; it is possible that they might shoot off to, say, $(100, 300)$ before returning to $(4, 12)$. Let's investigate the maximum and minimum values of $f(t) = t^3 - 6t^2 + 11t$ on the interval $0 \le t \le 4$. We can do this on a graphing calculator or computer, or use single-variable calculus. We already know the values of f at the endpoints, namely 0 and 12. We'll look for local extrema:

$$0 = f'(t) = 3t^2 - 12t + 11$$

which has roots at $t = 2 \pm \frac{1}{\sqrt{3}}$. These are the values of t where the particle changes direction: $t = 2 - \frac{1}{\sqrt{3}}$ corresponds to point P and $t = 2 + \frac{1}{\sqrt{3}}$ corresponds to point Q of C'. At these values of t we have $f(2 - \frac{1}{\sqrt{3}}) \approx 6.4$, and $f(2 + \frac{1}{\sqrt{3}}) \approx 5.6$. The fact that these values are between 0 and 12 shows that f takes on its maximum and minimum values at the endpoints of the interval and not in between.

(e) Using the parameterization given in part (d), we have

$$\vec{r}'(t) = x'(t)\vec{i} + y'(t)\vec{j} = \frac{1}{3}(3t^2 - 12t + 11)\vec{i} + (3t^2 - 12t + 11)\vec{j}.$$

Thus,

$$\int_{C'} \vec{F} \cdot d\vec{r}$$

$$= \int_0^4 \vec{F} \left(\frac{1}{3}(t^3 - 6t^2 + 11t), t^3 - 6t^2 + 11t \right) \cdot \left(\frac{1}{3}(3t^2 - 12t + 11)\vec{i} + (3t^2 - 12t + 11)\vec{j} \right) dt$$

$$= \int_0^4 \left(\frac{1}{3}(t^3 - 6t^2 + 11t)^2 \vec{i} + \frac{1}{3}(t^3 - 6t^2 + 11t)\vec{j} \right) \cdot \left(\frac{1}{3}(3t^2 - 12t + 11)\vec{i} + (3t^2 - 12t + 11)\vec{j} \right) dt$$

$$= \int_0^4 \frac{1}{3}(t^3 - 6t^2 + 11t)(3t^2 - 12t + 11) \left\{ ((t^3 - 6t^2 + 11t)\vec{i} + \vec{j}) \cdot (\frac{1}{3}\vec{i} + \vec{j}) \right\} dt$$

$$= \int_0^4 \frac{1}{3}(t^3 - 6t^2 + 11t)(3t^2 - 12t + 11) \left\{ \frac{1}{3}(t^3 - 6t^2 + 11t) + 1 \right\} dt$$

$$= \frac{1}{9} \int_0^4 (t^3 - 6t^2 + 11t)(3t^2 - 12t + 11)(t^3 - 6t^2 + 11t + 3) dt$$

Numerical integration yields an answer of 88, which agrees with the answer found in part b).

37. The integral corresponding to $A(t) = (t, t)$ is

$$\int_0^1 3t\, dt.$$

The integral corresponding to $D(t) = (e^t - 1, e^t - 1)$ is

$$3\int_0^{\ln 2} (e^{2t} - e^t)\, dt.$$

The substitution $s = e^t - 1$ has $ds = e^t\, dt$. Also $s = 0$ when $t = 0$ and $s = 1$ when $t = \ln 2$. Thus, substituting into the integral corresponding to $D(t)$ and using the fact that $e^{2t} = e^t \cdot e^t$ gives

$$3\int_0^{\ln 2} (e^{2t} - e^t)\, dt = 3\int_0^{\ln 2} (e^t - 1)e^t\, dt = \int_0^1 3s\, ds.$$

The integral on the right-hand side is the same as the integral corresponding to $A(t)$. Therefore we have

$$3\int_0^{\ln 2} (e^{2t} - e^t)\, dt = \int_0^1 3s\, ds = \int_0^1 3t\, dt.$$

Alternatively, the substitution $t = e^w - 1$ converts the integral corresponding to $A(t)$ into the integral corresponding to $B(t)$.

Solutions for Section 18.3

Exercises

1. Since $\vec{F}$ is a gradient field, with $\vec{F} = \operatorname{grad} f$ where $f(x, y) = x^2 + y^4$, we use the Fundamental Theorem of Line Integrals. The starting point of the path C is $(2, 0)$ and the end is $(0, 2)$. Thus,

$$\int_C \vec{F} \cdot d\vec{r} = f(0, 2) - f(2, 0) = 16 - 4 = 12.$$

5. Path-independent, because the vector field appears constant.

9. Since $\vec{F} = 3x^2\vec{i} + 4y^3\vec{j} = \operatorname{grad}(x^3 + y^4)$, we take $f(x, y) = x^3 + y^4$. Then by the Fundamental Theorem of Line Integrals,

$$\int_C \vec{F} \cdot d\vec{r} = f(-1, 0) - f(1, 0) = (-1)^3 - 1^3 = -2.$$

13. Since $\vec{F} = y\sin(xy)\vec{i} + x\sin(xy)\vec{j} = \operatorname{grad}(-\cos(xy))$, the Fundamental Theorem of Line Integrals gives

$$\int_C \vec{F} \cdot d\vec{r} = -\cos(xy)\Big|_{(1,2)}^{(3,18)} = -\cos(54) + \cos(2) = \cos(2) - \cos(54).$$

17. Since $\vec{F} = 2xy^2ze^{x^2y^2z}\vec{i} + 2x^2yze^{x^2y^2z}\vec{j} + x^2y^2e^{x^2y^2z}\vec{k} = \operatorname{grad}(e^{x^2y^2z})$ and the curve C is closed, the Fundamental Theorem of Line Integrals tells us that $\int_C \vec{F} \cdot d\vec{r} = 0$, since

$$\int_C \vec{F} \cdot d\vec{r} = e^{x^2y^2z}\Big|_{(1,0,1)}^{(1,0,1)} = e^0 - e^0 = 0.$$

Problems

21. Yes. If $f(x, y) = \frac{1}{2}x^2$, then grad $f = x\vec{i}$.

25. (a) To find the change in f by computing a line integral, we first choose a path C between the points; the simplest is a line. We parameterize the line by $(x(t), y(t)) = (t, \pi t/2)$, with $0 \leq t \leq 1$. Then $(x'(t), y'(t)) = (1, \pi/2)$, so the Fundamental Theorem of Line Integrals tells us that

$$
f(1, \frac{\pi}{2}) - f(0, 0) = \int_C \text{grad } f \cdot d\vec{r}
$$
$$
= \int_0^1 \text{grad } f\left(t, \frac{\pi t}{2}\right) \cdot \left(\vec{i} + \frac{\pi}{2}\vec{j}\right) dt
$$
$$
= \int_0^1 \left(2te^{t^2} \sin\left(\frac{\pi t}{2}\right)\vec{i} + e^{t^2} \cos\left(\frac{\pi t}{2}\right)\vec{j}\right) \cdot \left(\vec{i} + \frac{\pi}{2}\vec{j}\right) dt
$$
$$
= \int_0^1 \left(2te^{t^2} \sin\left(\frac{\pi t}{2}\right) + \frac{\pi e^{t^2}}{2} \cos\left(\frac{\pi t}{2}\right)\right) dt
$$
$$
= \int_0^1 \frac{d}{dt}\left(e^{t^2} \sin\left(\frac{\pi t}{2}\right)\right) dt
$$
$$
= e^{t^2} \sin\left(\frac{\pi t}{2}\right)\Big|_0^1 = e = 2.718.
$$

This integral can also be approximated numerically.

(b) The other way to find the change in f between these two points is to first find f. To do this, observe that

$$
2xe^{x^2} \sin y\vec{i} + e^{x^2} \cos y\vec{j} = \frac{\partial}{\partial x}\left(e^{x^2} \sin y\right)\vec{i} + \frac{\partial}{\partial y}\left(e^{x^2} \sin y\right)\vec{j} = \text{grad}\left(e^{x^2} \sin y\right).
$$

So one possibility for f is $f(x, y) = e^{x^2} \sin y$. Thus,

$$
\text{Change in } f\Big|_{(0,0)}^{(1,\pi/2)} = e^{x^2} \sin y\Big|_{(0,0)}^{(1,\pi/2)} = e^1 \sin\left(\frac{\pi}{2}\right) - e^0 \sin 0 = e.
$$

The exact answer confirms our calculations in part (a) which show that the answer is e.

29. This vector field is not a gradient field, so we evaluate the line integral directly. Let C_1 be the path along the x-axis from $(0, 0)$ to $(3, 0)$ and let C_2 be the path from $(3, 0)$ to $(3/\sqrt{2}, 3/\sqrt{2})$ along $x^2 + y^2 = 9$. Then

$$
\int_C \vec{H} \cdot d\vec{r} = \int_{C_1} \vec{H} \cdot d\vec{r} + \int_{C_2} \vec{H} \cdot d\vec{r}.
$$

On C_1, the vector field has only a $\vec{j}$ component (since $y = 0$), and $\vec{H}$ is therefore perpendicular to the path. Thus,

$$
\int_{C_1} \vec{H} \cdot d\vec{r} = 0.
$$

On C_2, the vector field is tangent to the path. The path is one eighth of a circle of radius 3 and so has length $2\pi(3/8) = 3\pi/4$.

$$
\int_{C_2} \vec{H} \cdot d\vec{r} = \|\vec{H}\| \cdot \text{Length of path} = 3 \cdot \left(\frac{3\pi}{4}\right) = \frac{9\pi}{4}.
$$

Thus,

$$
\int_C \vec{H} \cdot d\vec{r} = \frac{9\pi}{4}.
$$

33. Although this curve is complicated, the vector field is a gradient field since

$$
\vec{F} = \sin\left(\frac{x}{2}\right) \sin\left(\frac{y}{2}\right)\vec{i} - \cos\left(\frac{x}{2}\right) \cos\left(\frac{y}{2}\right)\vec{j} = \text{grad}\left(-2\cos\left(\frac{x}{2}\right) \sin\left(\frac{y}{2}\right)\right).
$$

Thus, only the endpoints of the curve, P and Q, are needed. Since $P = (-3\pi/2, 3\pi/2)$ and $Q = (-3\pi/2, -3\pi/2)$ and $\vec{F} = \text{grad}(-2\cos(x/2)\sin(y/2))$, we have

$$\int_C \vec{F} \cdot d\vec{r} = -2\cos\left(\frac{x}{2}\right)\sin\left(\frac{y}{2}\right)\Bigg|_{P=(-3\pi/2,3\pi/2)}^{Q=(-3\pi/2,-3\pi/2)}$$

$$= -2\cos\left(-\frac{3\pi}{4}\right)\sin\left(-\frac{3\pi}{4}\right) + 2\cos\left(-\frac{3\pi}{4}\right)\sin\left(\frac{3\pi}{4}\right)$$

$$= 2\cos\left(\frac{3\pi}{4}\right)\sin\left(\frac{3\pi}{4}\right) + 2\cos\left(\frac{3\pi}{4}\right)\sin\left(\frac{3\pi}{4}\right)$$

$$= -2 \cdot \frac{1}{\sqrt{2}} \cdot \frac{1}{\sqrt{2}} - 2 \cdot \frac{1}{\sqrt{2}} \cdot \frac{1}{\sqrt{2}} = -2.$$

37. (a) By the Fundamental Theorem of Line Integrals

$$\int_{(0,2)}^{(3,4)} \text{grad } f \cdot d\vec{r} = f(3,4) - f(0,2) = 66 - 57 = 9.$$

(b) By the Fundamental Theorem of Line Integrals, since C is a closed path, $\int_C \text{grad } f \cdot d\vec{r} = 0$.

41. (a) Work done by the force is the line integral, so

$$\text{Work done against force} = -\int_C \vec{F} \cdot d\vec{r} = -\int_C (-mg\,\vec{k}) \cdot d\vec{r}.$$

Since $\vec{r} = (\cos t)\vec{i} + (\sin t)\vec{j} + t\vec{k}$, we have $\vec{r}\,' = -(\sin t)\vec{i} + (\cos t)\vec{j} + \vec{k}$,

$$\text{Work done against force} = \int_0^{2\pi} mg\,\vec{k} \cdot (-\sin t\,\vec{i} + \cos t\,\vec{j} + \vec{k})dt$$

$$= \int_0^{2\pi} mg\,dt = 2\pi\,mg.$$

(b) We know from physical principles that the force is conservative. (Because the work done depends only on the vertical distance moved, not on the path taken.) Alternatively, we see that

$$\vec{F} = -mg\,\vec{k} = \text{grad}(-mgz),$$

so $\vec{F}$ is a gradient field and therefore path independent, or conservative.

45. (a) We have

$$\text{grad } h = \vec{i}$$
$$\text{grad } \phi = 2y\vec{i} + 2x\vec{j}$$
$$\vec{F} - \text{grad } \phi = -y\vec{i} = -y\,\text{grad } h.$$

Thus, $\vec{F} - \text{grad } \phi$ is a multiple of grad h.

(b) By part (a) the vector fields $\vec{F}$ and grad ϕ have the same components perpendicular to grad h, which is to say the same components in the direction of the level curve C of h. Thus, the line integrals of $\vec{F}$ and grad ϕ along C are equal. Using the Fundamental Theorem of Calculus for Line Integrals, we have

$$\int_C \vec{F} \cdot d\vec{r} = \int_C \text{grad } \phi \cdot d\vec{r} = \phi(Q) - \phi(P) = 60 - 30 = 30.$$

49. (a) By the chain rule

$$\frac{dh}{dt} = \frac{\partial f}{\partial x}\frac{dx}{dt} + \frac{\partial f}{\partial y}\frac{dy}{dt} = f_x x'(t) + f_y y'(t),$$

which is the result we want.

(b) Using the parameterization of C that we were given,

$$\int_C \operatorname{grad} f \cdot d\vec{r} = \int_a^b (f_x(x(t), y(t))\vec{i} + f_y(x(t), y(t))\vec{j}) \cdot (x'(t)\vec{i} + y'(t)\vec{j})dt$$

$$= \int_a^b (f_x(x(t), y(t))x'(t) + f_y(x(t), y(t))y'(t))dt.$$

Using the result of part (a), this gives us

$$\int_C \operatorname{grad} f \cdot d\vec{r} = \int_a^b h'(t)dt$$
$$= h(b) - h(a) = f(Q) - f(P).$$

Solutions for Section 18.4

Exercises

1. We know that

$$\frac{\partial f}{\partial x} = 2xy \quad \text{and} \quad \frac{\partial f}{\partial y} = x^2,$$

so, integrating with respect to x, thinking of y as a constant gives

$$f(x, y) = x^2 y + C(y).$$

Differentiating with respect to y gives

$$\frac{\partial f}{\partial y} = x^2 + C'(y),$$

so we take $C(y) = k$ for some constant K^2. Thus

$$f(x, y) = x^2 y + K.$$

5. Yes, since $\vec{F} = 2xy\vec{i} + x^2\vec{j} = \operatorname{grad}(x^2 y)$.

9. The domain of the vector field $\vec{F} = (2xy^3 + y)\vec{i} + (3x^2 y^2 + x)\vec{j}$ is the whole xy-plane. We apply the curl test:

$$\frac{\partial F_1}{\partial y} = 6xy^2 + 1 = \frac{\partial F_2}{\partial x}$$

so $\vec{F}$ is the gradient of a function f. In order to compute f we first integrate

$$\frac{\partial f}{\partial x} = 2xy^3 + y$$

with respect to x thinking of y as a constant. We get

$$f(x, y) = x^2 y^3 + xy + C(y)$$

Differentiating with respect to y and using the fact that $\partial f/\partial y = 3x^2 y^2 + x$ gives

$$\frac{\partial f}{\partial y} = 3x^2 y^2 + x + C'(y) = 3x^2 y^2 + x$$

Thus $C'(y) = 0$ so C is constant and

$$f(x, y) = x^2 y^3 + xy + C.$$

13. We have

$$\frac{\partial F_1}{\partial y} = \frac{(x^2 + y^2)1 - y(2y)}{(x^2 + y^2)^2} = \frac{x^2 - y^2}{(x^2 + y^2)^2}$$

$$\frac{\partial F_2}{\partial x} = -\frac{(x^2 + y^2)1 - x(2x)}{(x^2 + y^2)^2} = -\frac{y^2 - x^2}{(x^2 + y^2)^2} = \frac{x^2 - y^2}{(x^2 + y^2)^2}.$$

Thus $\dfrac{\partial F_1}{\partial y} = \dfrac{\partial F_2}{\partial x}$. However, the domain of the vector field contains a "hole" at the origin, so the curl test does not apply. This is not a gradient field. See Example 7 on page 957 of the text.

17. By Green's Theorem, with R representing the interior of the circle,

$$\int_C \vec{F} \cdot d\vec{r} = \int_R \left(\frac{\partial}{\partial x}(xy) - \frac{\partial}{\partial y}(3y) \right) dA = \int_R (y - 3)\, dA.$$

The integral of y over the interior of the circle is 0, by symmetry, because positive contributions of y from the top half of the circle cancel those from the bottom half. Thus

$$\int_R y\, dA = 0.$$

So

$$\int_C \vec{F} \cdot d\vec{r} = \int_R (y - 3)\, dA = \int_R -3\, dA = -3 \cdot \text{Area of circle} = -3 \cdot \pi(1)^2 = -3\pi.$$

Problems

21. The curve is closed, so we can use Green's Theorem. If R represents the interior of the region

$$\int_C \vec{F} \cdot d\vec{r} = \int_R \left(\frac{\partial F_2}{\partial x} - \frac{\partial F_1}{\partial y} \right) dA = \int_R \left(\frac{\partial(x)}{\partial x} - \frac{\partial(x - y)}{\partial y} \right) dA$$

$$= \int_R (1 - (-1))\, dA = \int_R 2\, dA = 2 \cdot \text{Area of sector}.$$

Since R is $1/8$ of a circle, R has area $\pi(3^2)/8$. Thus,

$$\int_C \vec{F} \cdot d\vec{r} = 2 \cdot \frac{9\pi}{8} = \frac{9\pi}{4}.$$

25. Since $\vec{F} = x\vec{j}$, we have $\partial F_2/\partial x = 1$ and $\partial F_1/\partial y = 0$. Thus, using Green's Theorem if R is the region enclosed by the closed curve C, we have

$$\int_C \vec{F} \cdot d\vec{r} = \int_R \left(\frac{\partial F_2}{\partial x} - \frac{\partial F_1}{\partial y} \right) dx\, dy = \int_R 1\, dx\, dy = \text{Area of } R$$

29. (a) The curve, C, is closed and oriented in the correct direction for Green's Theorem. See Figure 18.7. Writing R for the interior of the circle, we have

$$\int_C \left((x^2 - y)\vec{i} + (y^2 + x)\vec{j} \right) \cdot d\vec{r} = \int_R \left(\frac{\partial(y^2 + x)}{\partial x} - \frac{\partial(x^2 - y)}{\partial y} \right) dx\, dy$$

$$= \int_R (1 - (-1))\, dx\, dy = 2 \int_R dx\, dy$$

$$= 2 \cdot \text{Area of circle} = 2(\pi \cdot 3^2) = 18\pi.$$

(b) The circle given has radius R and center (a, b). The argument in part (a) works for any circle of radius R, oriented counterclockwise. So the line integral has the value $2\pi R^2$.

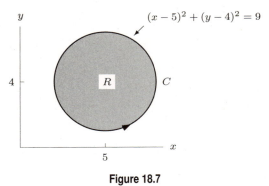

Figure 18.7

33. (a) We see that $\vec{F}, \vec{G}, \vec{H}$ are all gradient vector fields, since

$$\text{grad}(xy) = \vec{F} \quad \text{for all } x, y$$
$$\text{grad}(\arctan(x/y)) = \vec{G} \quad \text{except where } y = 0$$
$$\text{grad}\left((x^2 + y^2)^{1/2}\right) = \vec{H} \quad \text{except at } (0,0).$$

Other answer are possible. For example $\text{grad}(-\arctan(y/x)) = \vec{G}$ for $x \neq 0$.

(b) Parameterizing the unit circle, C, by $x = \cos t, y = \sin t, 0 \leq t \leq \pi$, we have $\vec{r}\,'(t) = -\sin t\vec{i} + \cos t\vec{j}$, so

$$\int_C \vec{F} \cdot d\vec{r} = \int_0^{2\pi} ((\sin t)\vec{i} + (\cos t)\vec{j}\,) \cdot ((-\sin t)\vec{i} + (\cos t)\vec{j}\,)\, dt = \int_0^{2\pi} \cos(2t)\, dt = 0.$$

The vector field $\vec{G}$ is tangent to the circle, pointing in the opposite direction to the parameterization, and of length 1 everywhere. Thus

$$\int_C \vec{G} \cdot d\vec{r} = -1 \cdot \text{ Length of circle } = -2\pi.$$

The vector field $\vec{H}$ points radially outward, so it is perpendicular to the circle everywhere. Thus

$$\int_C \vec{H} \cdot d\vec{r} = 0.$$

(c) Green's Theorem does not apply to the computation of the line integrals for $\vec{G}$ and $\vec{H}$ because their domains do not include the origin, which is in the interior, R, of the circle. Green's Theorem does apply to $\vec{F} = y\vec{i} + x\vec{j}$.

$$\int_C \vec{F} \cdot d\vec{r} = \int_R \left(\frac{\partial F_2}{\partial x} - \frac{\partial F_1}{\partial y}\right) dx\, dy = \int_R 0\, dx\, dy = 0.$$

37. (a) We use Green's Theorem. Let R be the region enclosed by the circle C. Then

$$\int_C \vec{F} \cdot d\vec{r} = \int_R \left(\frac{\partial F_2}{\partial x} - \frac{\partial F_1}{\partial y}\right) dA = \int_R \left(\frac{\partial}{\partial x}(e^{y^2} + 12x) - \frac{\partial}{\partial y}(3x^2 y + y^3 + e^x)\right) dA$$
$$= \int_R (12 - (3x^2 + 3y^2))\, dA = \int_R (12 - 3(x^2 + y^2))\, dA.$$

Converting to polar coordinates, we have

$$\int_C \vec{F} \cdot d\vec{r} = \int_0^{2\pi} \int_0^1 (12 - 3r^2)r\, dr\, d\theta = 2\pi \left(6r^2 - \frac{3}{4}r^4\right)\Big|_0^1 = 2\pi\left(6 - \frac{3}{4}\right) = \frac{21\pi}{2}.$$

(b) The integrand of the integral over the disk R is $12 - 3(x^2 + y^2)$. Since the integrand is positive for $x^2 + y^2 < 4$ and negative for $x^2 + y^2 > 4$, the integrand is positive inside the circle of radius 2 and negative outside that circle. Thus, the integral over R increases with a until $a = 2$ and then decreases. The maximum value of the line integral occurs when $a = 2$.

Solutions for Chapter 18 Review

Exercises

1. The angle between the vector field and the curve is more than $90°$ at all points on C, so the line integral is negative.

5. Since $\vec{F} = 6\vec{i} - 7\vec{j}$, consider the function f
$$f(x, y) = 6x - 7y.$$
Then we see that grad $f = 6\vec{i} - 7\vec{j}$, so we use the Fundamental Theorem of Calculus for Line Integrals:
$$\int_C \vec{F} \cdot d\vec{r} = \int_C \text{grad } f \cdot d\vec{r}$$
$$= f(4, 4) - f(2, -6) = (-4) - (54) = -58.$$

9. Only the $\vec{j}$ component contributes to the integral. On the y-axis, $x = 0$, so
$$\int_C \vec{F} \cdot d\vec{r} = \int_3^5 y^2\vec{j} \cdot \vec{j} \; dy = \frac{y^3}{3}\bigg|_3^5 = \frac{98}{3}.$$

13. The triangle C consists of the three paths shown in Figure 18.8.

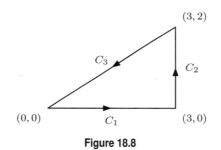

$(3, 2)$

C_3

C_2

$(0, 0)$

C_1

$(3, 0)$

Figure 18.8

Write $C = C_1 + C_2 + C_3$ where C_1, C_2, and C_3 are parameterized by
$$C_1 : (t, 0) \text{ for } 0 \leq t \leq 3; \quad C_2 : (3, t) \text{ for } 0 \leq t \leq 2; \quad C_3 : (3 - 3t, 2 - 2t) \text{ for } 0 \leq t \leq 1.$$
Then
$$\int_C \vec{F} \cdot d\vec{r} = \int_{C_1} \vec{F} \cdot d\vec{r} + \int_{C_2} \vec{F} \cdot d\vec{r} + \int_{C_3} \vec{F} \cdot d\vec{r}$$
where
$$\int_{C_1} \vec{F} \cdot d\vec{r} = \int_0^3 \vec{F}(t, 0) \cdot \vec{i} \; dt = \int_0^3 (2t + 4)dt = (t^2 + 4t)\big|_0^3 = 21$$
$$\int_{C_2} \vec{F} \cdot d\vec{r} = \int_0^2 \vec{F}(3, t) \cdot \vec{j} \; dt = \int_0^2 (5t + 3)dt = (5t^2/2 + 3t)\big|_0^2 = 16$$
$$\int_{C_3} \vec{F} \cdot d\vec{r} = \int_0^1 \vec{F}(3 - 3t, 2 - 2t) \cdot (-3\vec{i} - 2\vec{j})dt$$
$$= \int_0^1 ((-4t + 8)\vec{i} + (-19t + 13)\vec{j}) \cdot (-3\vec{i} - 2\vec{j})dt$$
$$= 50 \int_0^1 (t - 1)dt = -25.$$

So

$$\int_C \vec{F} \, d\vec{r} = 21 + 16 - 25 = 12.$$

17. The domain is all 3-space. Since $F_1 = y$,

$$\text{curl } y\vec{i} = \left(\frac{\partial F_3}{\partial y} - \frac{\partial F_2}{\partial z} \right) \vec{i} + \left(\frac{\partial F_1}{\partial z} - \frac{\partial F_3}{\partial x} \right) \vec{j} + \left(\frac{\partial F_2}{\partial x} - \frac{\partial F_1}{\partial y} \right) \vec{k} = -\vec{k} \neq \vec{0},$$

so $\vec{F}$ is not path-independent.

21. The domain is all 3-space. Since $F_1 = y$, $F_2 = x$,

$$\text{curl } y\vec{i} + x\vec{j} = \left(\frac{\partial F_3}{\partial y} - \frac{\partial F_2}{\partial z} \right) \vec{i} + \left(\frac{\partial F_1}{\partial z} - \frac{\partial F_3}{\partial x} \right) \vec{j} + \left(\frac{\partial F_2}{\partial x} - \frac{\partial F_1}{\partial y} \right) \vec{k} = \vec{0},$$

so $\vec{F}$ is path-independent

25. Since the line is parallel to the x-axis, only the $\vec{i}$-component contributes to the line integral. On C, we have $d\vec{r} = \vec{i} \, dx$, so

$$\int_C \vec{F} \cdot d\vec{r} = \int_2^{12} 5x\vec{i} \cdot \vec{i} \, dx = \frac{5}{2} x^2 \bigg|_2^{12} = 350.$$

29. We can calculate this line integral either by calculating a separate line integral for each side, or by adding a line segment, C_1, from $(1, 4)$ to $(1, 1)$ to form the closed curve $C + C_1$. Since we now have a closed curve, we can use Green's Theorem:

$$\int_{C+C_1} \vec{F} \cdot d\vec{r} = \int_{C+C_1} (5x\vec{i} + 3x\vec{j}) \cdot d\vec{r} = \int_R \left(\frac{\partial}{\partial x}(3x) - \frac{\partial}{\partial y}(5y) \right) dx \, dy$$

$$= \int_R 3 \, dx \, dy = 3 \cdot \text{Area of region} = 3 \left(2 \cdot 3 + \frac{1}{2} 3 \cdot 4 \right) = 36.$$

Since $d\vec{r} = -\vec{j} \, dy$ on C_1, we have

$$\int_{C_1} \vec{F} \cdot d\vec{r} = \int_4^1 3 \cdot 1\vec{j} \cdot (-\vec{j} \, dy) = -3 \cdot 3 = -9.$$

Since

$$\int_{C+C_1} \vec{F} \cdot d\vec{r} = \int_C \vec{F} \cdot d\vec{r} + \int_{C_1} \vec{F} \cdot d\vec{r} = \int_C \vec{F} \cdot d\vec{r} - 9 = 36$$

we have

$$\int_C \vec{F} \cdot d\vec{r} = 45.$$

Problems

33. (a) The vector field is everywhere perpendicular to the radial line from the origin to $(2, 3)$, so the line integral is 0.

(b) Since the path is parallel to the x-axis, only the $\vec{i}$ component of the vector field contributes to the line integral. The $\vec{i}$ component is $-3\vec{i}$ on this line, and the displacement along this line is $-2\vec{i}$, so

$$\text{Line integral} = (-3\vec{i}) \cdot (-2\vec{i}) = 6.$$

(c) The circle of radius 5 has equation $x^2 + y^2 = 25$. On this curve, $\|\vec{F}\| = \sqrt{(-y^2) + x^2} = \sqrt{25} = 5$. In addition, $\vec{F}$ is everywhere tangent to the circle, and the path is $3/4$ of the circle. Thus

$$\text{Line integral} = \|\vec{F}\| \cdot \text{Length of curve} = 5 \cdot \frac{3}{4} \cdot 2\pi(5) = \frac{75}{2}\pi.$$

(d) Use Green's Theorem. Writing C for the curve around the boundary of the triangle, we have

$$\frac{\partial F_2}{\partial x} - \frac{\partial F_1}{\partial y} = 1 - (-1) = 2,$$

so

$$\int_C \vec{F} \cdot d\vec{r} = \int_{\text{Triangle}} 2 \, dA = 2 \cdot \text{Area of triangle} = 2 \cdot 7 = 14.$$

37. (a) Since $\vec{F} = (6x + y^2)\vec{i} + 2xy\vec{j} = \operatorname{grad}(3x^2 + xy^2)$, the vector field $\vec{F}$ is path independent, so

$$\int_{C_1} \vec{F} \cdot d\vec{r} = 0.$$

(b) Since C_1 is closed, we use Green's Theorem, so

$$\int_{C_1} \vec{G} \cdot d\vec{r} = \int_{\text{Interior of } C_1} \left(\frac{\partial}{\partial x}(x+y) - \frac{\partial}{\partial y}(x-y) \right) dA$$

$$= 2 \int_{C_1} dA = 2 \cdot \text{Area inside } C_1 = 2 \cdot \frac{1}{2} \cdot 2 \cdot 2 = 4.$$

(c) Since $\vec{F} = \operatorname{grad}(3x^2 + xy^2)$, using the Fundamental Theorem of Line Integrals gives

$$\int_{C_2} \vec{F} \cdot d\vec{r} = (3x^2 + xy^2)\Big|_{(2,0)}^{(0,-2)} = 0 - 3 \cdot 2^2 = -12.$$

(d) Parameterizing the circle by

$$x = 2\cos t \qquad y = 2\sin t \qquad 0 \le t \le \frac{3\pi}{2},$$

gives

$$x' = -2\sin t \qquad y' = 2\cos t,$$

so the integral is

$$\int_{C_2} \vec{G} \cdot d\vec{r} = \int_0^{3\pi/2} \left((2\cos t - 2\sin t)\vec{i} + (2\cos t + 2\sin t)\vec{j} \right) \cdot (-2\sin t\,\vec{i} + 2\cos t\,\vec{j})\, dt$$

$$= \int_0^{3\pi/2} 4(-\cos t \sin t + \sin^2 t + \cos^2 t + \sin t \cos t)\, dt$$

$$= 4 \int_0^{3\pi/2} dt = 4 \cdot \frac{3\pi}{2} = 6\pi.$$

41. (a) Since $\dfrac{\partial}{\partial x}(y^5 + x) - \dfrac{\partial}{\partial y}(x^3 - y) = 1 + 1 = 2$, any closed curve oriented counterclockwise will do. See Figure 18.9.

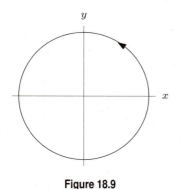

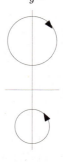

Figure 18.9   **Figure 18.10**

(b) Since $\dfrac{\partial}{\partial x}(y^5 - xy) - \dfrac{\partial}{\partial y}(x^3) = -y$, any closed curve in the lower half-plane oriented counterclockwise or any closed curve in the upper half-plane oriented clockwise will do. See Figure 18.10. Other answers are possible.

45. Since $\|\vec{F}\| \le 7$, the line integral cannot be larger than 7 times the length of the curve. Thus

$$\int_C \vec{F} \cdot d\vec{r} \le 7 \cdot \text{Circumference of circle} = 7 \cdot 2\pi = 14\pi.$$

The line integral is equal to 14π if $\vec{F}$ is everywhere of magnitude 7, tangent to the curve, and pointing in the direction in which the curve is traversed.

The smallest possible value occurs if the vector field is everywhere of magnitude 7, tangent to the curve and pointing opposite to the direction in which the curve is transversed. Thus

$$\int_C \vec{F} \cdot d\vec{r} \geq -14\pi.$$

49. The free vortex appears to starts at about $r = 200$ meters (that's where the graph changes its behavior) and the tangential velocity at this point is about 200 km/hr $= 2 \cdot 10^5$ meters/hr.

Since $\vec{v} = \omega(-y\vec{i} + x\vec{j})$ for $\sqrt{x^2 + y^2} \leq 200$, at $r = 200$ we have

$$\|\vec{v}\| = \omega\sqrt{(-y)^2 + x^2} = \omega(200) = 2 \cdot 10^5 \text{ meters/hr},$$

so

$$\omega = 10^3 \text{ rad/hr}.$$

Since $\vec{v} = K(x^2 + y^2)^{-1}(-y\vec{i} + x\vec{j})$ for $\sqrt{x^2 + y^2} \geq 200$, at $r = 200$ we have

$$\|\vec{v}\| = K(200^2)^{-1}(200) = \frac{K}{200} = 2 \cdot 10^5 \text{ meters/hr}$$

so

$$K = 4 \cdot 10^7 \text{ m}^2 \cdot \text{rad/hr}.$$

CAS Challenge Problems

53. We have

$$\int_{C_1} \vec{F} \cdot d\vec{r} = \int_0^3 \vec{F}(\vec{r}(t)) \cdot \vec{r}'(t) dt$$

$$= \int_0^3 \left(2\left(2at + bt^2\right) + 2t\left(2ct + dt^2\right)\right) dt = 18a + 18b + 36c + (81d/2)$$

and

$$\int_{C_2} \vec{F} \cdot d\vec{r} = \int_0^3 \vec{F}(\vec{r}(t)) \cdot \vec{r}'(t) dt$$

$$= \int_0^3 \left(-2\left(2a(3-t) + b(3-t)^2\right) - 2\left(2c(3-t) + d(3-t)^2\right)(3-t)\right) dt$$

$$= -18a - 18b - 36c - (81d/2).$$

The second integral is the negative of the first. This is because C_2 is the same curve as C_1 but traveling in the opposite direction.

CHECK YOUR UNDERSTANDING

1. A path-independent vector field must have zero circulation around all closed paths. Consider a vector field like $\vec{F}(x, y) = |x|\vec{j}$, shown in Figure 18.11.

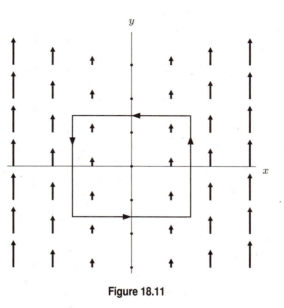

Figure 18.11

A rectangular path that is symmetric about the y-axis will have zero circulation: on the horizontal sides, the field is perpendicular, so the line integral is zero. The line integrals on the vertical sides are equal in magnitude and opposite in sign, so they cancel out, giving a line integral of zero. However, this field is not path-independent, because it is possible to find two paths with the same endpoints but different values of the line integral of $\vec{F}$. For example, consider the two points $(0, 0)$ and $(0, 1)$. The path C_1 in Figure 18.12 along the y axis gives zero for the line integral, because the field is 0 along the y axis, whereas a path like C_2 will have a nonzero line integral. Thus the line integral depends on the path between the points, so $\vec{F}$ is not path-independent.

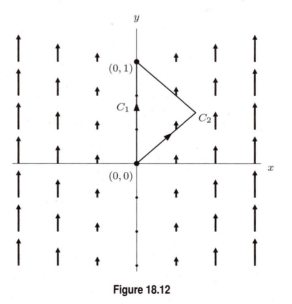

Figure 18.12

5. False. Because $\vec{F} \cdot \Delta \vec{r}$ is a scalar quantity, $\int_C \vec{F} \cdot d\vec{r}$ is also a scalar quantity.

9. False. We can calculate a line integral of any vector field.

13. False. The relative sizes of the line integrals along C_1 and C_2 depend on the behavior of the vector field $\vec{F}$ along the curves. As a counterexample, take the vector field $\vec{F} = \vec{i}$, and C_1 to be the line from the origin to $(0, 2)$, while C_2 is the line from the origin to $(1, 0)$. Then the length of C_1 is 2, which is greater than the length of C_2, which is 1. However $\int_{C_1} \vec{F} \cdot d\vec{r} = 0$ (since $\vec{F}$ is perpendicular to C_1) while $\int_{C_2} \vec{F} \cdot d\vec{r} > 0$ (since $\vec{F}$ points along C_2).

17. False. The relation between these two line integrals depends on the behavior of the vector field along each of the curves, so there is no reason to expect one to be the negative of the other. As an example, if $\vec{F}(x, y) = y\vec{i}$, then, by symmetry, both line integrals are equal to the same negative number.

21. True. The dot product of the integrand $4\vec{i}$ with $\vec{r}'(t) = \vec{i} + 2t\vec{j}$ is 4, so the integral has value $\int_0^2 4 \, dt = 8$.

25. False. As a counterexample, consider the vector field $\vec{F} = x\vec{i}$. Then if we parameterize C_1 by $\vec{r}(t) = t\vec{i}$, with $0 \leq t \leq 1$, we get

$$\int_{C_1} x\vec{i} \cdot d\vec{r} = \int_0^1 t\vec{i} \cdot \vec{i} \, dt = \int_0^1 t \, dt = \left. \frac{t^2}{2} \right|_0^1 = \frac{1}{2}.$$

A similar computation for C_2 gives a line integral with value 2.

29. True. By the Fundamental Theorem for Line Integrals, if C is a path from P to Q, then $\int_C \operatorname{grad} f \cdot d\vec{r} = f(Q) - f(P)$, so the value of the line integral $\int_C \operatorname{grad} f \cdot d\vec{r}$ depends only on the endpoints and not the path.

33. True. Since a gradient field is path-independent, and C_1 and C_2 have the same initial and final points, the two line integrals are equal.

37. True. The value of $\dfrac{\partial F_2}{\partial x} - \dfrac{\partial F_1}{\partial y}$ is $0 - 0 = 0$, so the field is path-independent.

41. True. This vector field has components $F_1 = x$, $F_2 = y$, and $F_3 = z$. Using the 3-space curl test gives zero for all of the components of $\operatorname{curl} \vec{F}$, so the field is path-independent.

CHAPTER NINETEEN

Solutions for Section 19.1

Exercises

1. Scalar. Only the $\vec{j}$-component of the vector field contributes to the flux and $d\vec{A} = -\vec{j}\, dA$, so

$$\int_S (3\vec{i} + 4\vec{j}) \cdot d\vec{A} = -4 \cdot \text{Area of disk} = -4 \cdot \pi 5^2 = -100\pi.$$

5. On the surface, $d\vec{A} = \vec{k}\, dA$, so only the $\vec{k}$ component of $\vec{v}$ contributes to the flux:

$$\text{Flux} = \int_S \vec{v} \cdot d\vec{A} = \int_S (\vec{i} - \vec{j} + 3\vec{k}) \cdot \vec{k}\, dA = 3 \cdot \text{Area of disk} = 3 \cdot \pi 2^2 = 12\pi.$$

9. $\vec{v} \cdot \vec{A} = (2\vec{i} + 3\vec{j} + 5\vec{k}) \cdot \vec{k} = 5.$

13. The disk has area 25π, so its area vector is $25\pi\vec{j}$. Thus

$$\text{Flux} = (2\vec{i} + 3\vec{j}) \cdot 25\pi\vec{j} = 75\pi.$$

17. Since the square, S, is in the plane $y = 0$ and oriented in the negative y-direction, $d\vec{A} = -\vec{j}\, dxdz$ and

$$\int_S \vec{F} \cdot d\vec{A} = \int_S (0 + 3)\vec{j} \cdot (-\vec{j}\, dxdz) = -3 \int_S dxdz = -3 \cdot \text{Area of square} = -3(2^2) = -12.$$

21. The only contribution to the flux is from the $\vec{j}$-component, and since $d\vec{A} = \vec{j}\, dx\, dz$ on the square, S, we have

$$\text{Flux} = \int_S (6\vec{i} + x^2\vec{j} - \vec{k}) \cdot d\vec{A} = \int_{-2}^{2} \int_{-2}^{2} x^2\vec{j} \cdot \vec{j}\, dx\, dz = \int_{-2}^{2} \left. \frac{x^3}{3} \right|_{-2}^{2} dz = \frac{16}{3} \cdot 4 = \frac{64}{3}.$$

25. We have $d\vec{A} = \vec{i}\, dA$, so

$$\int_S \vec{F} \cdot d\vec{A} = \int_S (2z\vec{i} + x\vec{j} + x\vec{k}) \cdot \vec{i}\, dA = \int_S 2z\, dA$$

$$= \int_0^2 \int_0^3 2z\, dzdy = 18.$$

29. Since the disk is oriented in the positive x-direction, $d\vec{A} = \vec{i}\, dydz$, so we have

$$\text{Flux} = \int_{\text{Disk}} \vec{F} \cdot d\vec{A} = \int_{\text{Disk}} e^{y^2 + z^2}\vec{i} \cdot \vec{i}\, dydz = \int_{\text{Disk}} e^{y^2 + z^2}\, dydz.$$

To calculate this integral, we use polar coordinates with $y = r\cos\theta$ and $z = r\sin\theta$. Then $r^2 = y^2 + z^2$ and

$$\int_{\text{Disk}} \vec{F} \cdot d\vec{A} = \int_0^{2\pi} \int_0^2 e^{r^2} \cdot r dr d\theta = 2\pi \cdot \left. \frac{e^{r^2}}{2} \right|_0^2 = \pi(e^4 - 1).$$

33. Since $\vec{r}$ is perpendicular to S and $\|\vec{r}\| = 3$ on S, we have

$$\int_S \vec{r} \cdot d\vec{A} = 3 \cdot \text{Area of surface} = 3 \cdot 4\pi 3^2 = 108\pi.$$

Problems

37. Since this vector field points radially out from the origin, it is everywhere parallel to the vector representing the surface area, $d\vec{A}$. Thus since $\|\vec{F}(\vec{r})\| = 1/R^2$ on the surface, S,

$$\vec{F}(\vec{r}) \cdot d\vec{A} = \frac{1}{R^2}\, dA,$$

so

$$\int_S \vec{F}(\vec{r}) \cdot d\vec{A} = \frac{1}{R^2} \cdot \text{Surface area of sphere} = \frac{1}{R^2}(4\pi R^2) = 4\pi.$$

41. By the symmetry of the sphere, the $\vec{i}$ and $\vec{j}$ components of $\vec{F}$ do not contribute to the flux; only the $\vec{k}$ component contributes. The vector normal to S has a negative $\vec{k}$ component, so we need $c > 0$. There are no conditions on a and b.

45. The square of side 2 in the plane $x = 5$, oriented in the positive x-direction, has area vector $\vec{A} = 4\vec{i}$. Since the vector field is constant

$$\text{Flux} = (a\vec{i} + b\vec{j} + c\vec{k}) \cdot 4\vec{i} = 4a = 24.$$

Thus, $a = 6$ and we cannot say anything about the values of b and c.

49. (a)   (i) The integral $\int_W \rho\, dV$ represents the total charge in the volume W.

   (ii) The integral $\int_S \vec{J} \cdot d\vec{A}$ represents the total current flowing out of the surface S.

(b) The total current flowing out of the surface S is the rate at which the total charge inside the surface S (i.e., in the volume W) is decreasing. In other words,

$$\text{Rate current flowing out of } S = -\frac{\partial}{\partial t}(\text{charge in } W),$$

so

$$\int_S \vec{J} \cdot d\vec{A} = -\frac{\partial}{\partial t}\left(\int_W \rho\, dV\right).$$

Solutions for Section 19.2

Exercises

1. Only the z-component of the vector field contributes to the flux. Since $d\vec{A} = \vec{k}\, dx\, dy$ on the surface, we have

$$\int_S (3\vec{i} + 4\vec{j} + xy\vec{k}) \cdot d\vec{A} = \int_0^7 \int_0^5 xy\, dx\, dy = \int_0^7 \left.\frac{x^2 y}{2}\right|_0^5 dy = \int_0^7 \frac{25}{2} y\, dy = \left.\frac{25 y^2}{4}\right|_0^7 = \frac{1225}{4}.$$

5. Writing the surface S as $z = f(x,y) = -y + 1$, we have

$$d\vec{A} = (-f_x\vec{i} - f_y\vec{j} + \vec{k})dx\, dy.$$

Thus,

$$\int_S \vec{F} \cdot d\vec{A} = \int_R \vec{F}(x, y, f(x,y)) \cdot (-f_x\vec{i} - f_y\vec{j} + \vec{k})\, dx\, dy$$

$$= \int_0^1 \int_0^1 (2x\vec{j} + y\vec{k}) \cdot (\vec{j} + \vec{k})\, dx\, dy$$

$$= \int_0^1 \int_0^1 (2x + y)\, dx\, dy = \int_0^1 \left.(x^2 + xy)\right|_0^1 dy$$

$$= \int_0^1 (1 + y)\, dy = \left.(y + \frac{y^2}{2})\right|_0^1 = \frac{3}{2}.$$

9. On the curved side of the cylinder, only the components $x\vec{i} + z\vec{k}$ contribute to the flux. Since $x\vec{i} + z\vec{k}$ is perpendicular to the curved surface and $\|x\vec{i} + z\vec{k}\| = 2$ there (because the cylinder has radius 2), we have

$$\text{Flux through sides} = 2 \cdot \text{Area of curved surface} = 2 \cdot 2\pi \cdot 2 \cdot 6 = 48\pi.$$

On the flat ends, only $y\vec{j}$ contributes to the flux. On one end, $y = 3$ and $d\vec{A} = \vec{j}\, dA$; on the other end, $y = -3$ and $d\vec{A} = -\vec{j}\, dA$. Thus

$$\text{Flux through ends} = \text{Flux through top} + \text{Flux through bottom}$$
$$= 3\vec{j} \cdot \vec{j}\, \pi(2^2) + (-3\vec{j}) \cdot (-\vec{j}\,\pi(2^2)) = 24\pi.$$

So,

$$\text{Total flux} = 48\pi + 24\pi = 72\pi.$$

13. Using $z = 1 - x - y$, the upward pointing area element is $d\vec{A} = (\vec{i} + \vec{j} + \vec{k})\, dx\, dy$, so the downward one is $d\vec{A} = (-\vec{i} - \vec{j} - \vec{k})\, dx\, dy$. Since S is oriented downward, we have

$$\int_S \vec{F} \cdot d\vec{A} = \int_S (x\vec{i} + y\vec{j} + z\vec{k}) \cdot d\vec{A}$$
$$= \int_0^3 \int_0^2 (x\vec{i} + y\vec{j} + (1 - x - y)\vec{k}) \cdot (-\vec{i} - \vec{j} - \vec{k})\, dxdy$$
$$= \int_0^3 \int_0^2 (-x - y - 1 + x + y)\, dxdy = -6.$$

17. Since $y = f(x, z) = x^2 + z^2$, we have

$$d\vec{A} = (-f_x\vec{i} + \vec{j} - f_z\vec{k})\, dxdz = (-2x\vec{i} + \vec{j} - 2z\vec{k})\, dxdz.$$

Thus, substituting $y = x^2 + z^2$ into $\vec{F}$, we have

$$\int_S \vec{F} \cdot d\vec{A} = \int_{x^2+z^2 \leq 1} ((x^2 + z^2)\vec{i} + \vec{j} - xz\vec{k}) \cdot (-2x\vec{i} + \vec{j} - 2z\vec{k})\, dxdz$$
$$= \int_{x^2+z^2 \leq 1} (-2x^3 - 2xz^2 + 1 + 2xz^2)\, dxdz$$
$$= \int_{-1}^1 \int_{-\sqrt{1-z^2}}^{\sqrt{1-z^2}} (1 - 2x^3)\, dxdz$$
$$= \int_{-1}^1 \int_{-\sqrt{1-z^2}}^{\sqrt{1-z^2}} dxdz - \int_{-1}^1 \int_{-\sqrt{1-z^2}}^{\sqrt{1-z^2}} 2x^3\, dxdz$$
$$= \text{Area of disk} - \int_{-1}^1 \left(\left. \frac{x^4}{2} \right|_{-\sqrt{1-z^2}}^{\sqrt{1-z^2}} \right) dz = \pi - 0 = \pi$$

21. The flux of $\vec{F}$ through S is given by

$$\int_S \vec{F} \cdot d\vec{A} = \int_0^{2\pi} \int_0^{\pi/2} (2\cos\phi\vec{k}) \cdot (\sin\phi\cos\theta\vec{i} + \sin\phi\sin\theta\vec{j} + \cos\phi\vec{k})2^2 \sin\phi\, d\phi\, d\theta$$
$$= \int_{\theta=0}^{2\pi} \int_{\phi=0}^{\pi/2} 8\sin\phi\cos^2\phi\, d\phi d\theta = 16\pi \left(\frac{-\cos^3\phi}{3} \right) \Big|_{\phi=0}^{\pi/2} = \frac{16\pi}{3}.$$

Problems

25. The $\vec{k}$-component of $\vec{F}$ does not contribute to the flux as it is perpendicular to the surface. The vector field $x\vec{i} + y\vec{j}$ is everywhere perpendicular to S and has constant magnitude $||x^2 + y^2|| = 1$ on the surface S. Thus

$$\int_S \vec{F} \cdot d\vec{A} = \int_S (x\vec{i} + y\vec{j}) \cdot d\vec{A} = 1 \cdot \text{Area of } S = 1\frac{\pi}{2} = \frac{\pi}{2}.$$

Alternatively, the flux can be computed by integrating with respect to x and z, treating y as a function of x and z. A parameterization of S is given by $y = \sqrt{1 - x^2}, 0 \le x \le 1, 0 \le y \le 1, 0 \le z \le 1$. Thus,

$$\int_S \vec{F} \cdot d\vec{A} = \int_0^1 \int_0^1 (x\vec{i} + \sqrt{1 - x^2}\vec{j} + z\vec{k}) \cdot (-y_x\vec{i} + \vec{j} - y_z\vec{k})\, dx\, dz$$

$$= \int_0^1 \int_0^1 (x\vec{i} + \sqrt{1 - x^2}\vec{j} + z\vec{k}) \cdot \left(\frac{x}{\sqrt{1 - x^2}}\vec{i} + \vec{j} + 0\vec{k}\right) dx\, dz$$

$$= \int_0^1 \int_0^1 \left(\frac{x^2}{\sqrt{1 - x^2}} + \sqrt{1 - x^2}\right) dx\, dz$$

$$= \int_0^1 \int_0^1 \frac{1}{\sqrt{1 - x^2}}\, dx\, dz$$

$$= 1 \cdot \arcsin x \Big|_0^1 = \frac{\pi}{2}.$$

29. The plane is $x - z = 0$ over region $0 \le x \le \sqrt{2}, 0 \le y \le 2$. See Figure 19.1.

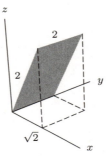

Figure 19.1

$$\text{Flux} = \int_0^2 \int_0^{\sqrt{2}} \left((e^{xy} + 3z + 5)\vec{i} + (e^{xy} + 5z + 3)\vec{j} + (3z + e^{xy})\vec{k}\right) \cdot (\vec{i} - \vec{k})\, dx\, dy$$

$$= \int_0^2 \int_0^{\sqrt{2}} (e^{xy} + 3z + 5 - 3z - e^{xy})\, dx\, dy = 5(2)(\sqrt{2}) = 10\sqrt{2}$$

Alternatively, since a unit normal to the surface is $\vec{n}/\sqrt{2} = (\vec{i} - \vec{j})/\sqrt{2}$, writing $dA = ||d\vec{A}||$, we have

$$\text{Flux} = \int_S \vec{H} \cdot d\vec{A} = \int \vec{H} \cdot \frac{\vec{i} - \vec{k}}{\sqrt{2}}\, dA = \int \frac{5}{\sqrt{2}}\, dA$$

$$= \frac{5}{\sqrt{2}}(\text{Area of slanted square}) = \frac{5}{\sqrt{2}}4 = 10\sqrt{2}.$$

Solutions for Section 19.3

Exercises

1. Since S is given by

$$\vec{r}(s,t) = (s+t)\vec{i} + (s-t)\vec{j} + (s^2+t^2)\vec{k},$$

we have

$$\frac{\partial \vec{r}}{\partial s} = \vec{i} + \vec{j} + 2s\vec{k} \quad \text{and} \quad \frac{\partial \vec{r}}{\partial t} = \vec{i} - \vec{j} + 2t\vec{k},$$

and

$$\frac{\partial \vec{r}}{\partial s} \times \frac{\partial \vec{r}}{\partial t} = \begin{vmatrix} \vec{i} & \vec{j} & \vec{k} \\ 1 & 1 & 2s \\ 1 & -1 & 2t \end{vmatrix} = (2s+2t)\vec{i} + (2s-2t)\vec{j} - 2\vec{k}.$$

Since the $\vec{i}$ component of this vector is positive for $0 < s < 1, 0 < t < 1$, it points away from the z-axis, and so has the opposite orientation to the one specified. Thus, we use

$$d\vec{A} = -\frac{\partial \vec{r}}{\partial s} \times \frac{\partial \vec{r}}{\partial t}\, ds\, dt,$$

and so we have

$$\int_S \vec{F} \cdot d\vec{A} = -\int_0^1 \int_0^1 (s^2+t^2)\vec{k} \cdot \left((2s+2t)\vec{i} + (2s-2t)\vec{j} - 2\vec{k} \right)\, ds\, dt$$

$$= 2\int_0^1 \int_0^1 (s^2+t^2)\, ds\, dt = 2\int_0^1 \left(\frac{s^3}{3} + st^2 \right) \Bigg|_{s=0}^{s=1} dt$$

$$= 2\int_0^1 \left(\frac{1}{3} + t^2 \right) dt = 2\left(\frac{1}{3}t + \frac{t^3}{3} \right) \Bigg|_0^1 = 2\left(\frac{1}{3} + \frac{1}{3} \right) = \frac{4}{3}.$$

5. The cross product $\partial \vec{r}/\partial s \times \partial \vec{r}/\partial t$ is given by

$$\frac{\partial \vec{r}}{\partial s} \times \frac{\partial \vec{r}}{\partial t} = \begin{vmatrix} \vec{i} & \vec{j} & \vec{k} \\ 2s & 2 & 0 \\ 0 & 2t & 5 \end{vmatrix} = 10\vec{i} - 10s\vec{j} + 4st\vec{k}.$$

Since the z-component, $4st$, of the vector $\partial \vec{r}/\partial s \times \partial \vec{r}/\partial t$ is positive for $0 < s \leq 1, 1 \leq t \leq 3$, we see that $\partial \vec{r}/\partial s \times \partial \vec{r}/\partial t$ points upward, in the direction of the orientation of S we were given. Thus, we use

$$d\vec{A} = \left(\frac{\partial \vec{r}}{\partial s} \times \frac{\partial \vec{r}}{\partial t} \right) ds\, dt,$$

and so we have

$$\int_S \vec{F} \cdot d\vec{A} = \int_0^1 \int_1^3 (5t\vec{i} + s^2\vec{j}) \cdot (10\vec{i} - 10s\vec{j} + 4st\vec{k})\, dt\, ds$$

$$= \int_0^1 \int_1^3 (50t - 10s^3)\, dt\, ds = \int_0^1 (25t^2 - 10s^3 t) \Bigg|_{t=1}^{t=3} ds$$

$$= \int_0^1 (200 - 20s^3)\, ds = (200s - 5s^4) \Bigg|_0^1$$

$$= 200 - 5 = 195.$$

9. A parameterization of S is

$$x = s, \quad y = t, \quad z = 3s + 2t, \quad \text{for } 0 \leq s \leq 10, \quad 0 \leq t \leq 20.$$

We compute

$$\frac{\partial \vec{r}}{\partial s} \times \frac{\partial \vec{r}}{\partial t} = (\vec{i} + 3\vec{k}) \times (\vec{j} + 2\vec{k}) = -3\vec{i} - 2\vec{j} + \vec{k}$$

$$\left\| \frac{\partial \vec{r}}{\partial \theta} \times \frac{\partial \vec{r}}{\partial t} \right\| = \sqrt{14}$$

$$\text{Surface area} = \int_S dA = \int_R \left\| \frac{\partial \vec{r}}{\partial \theta} \times \frac{\partial \vec{r}}{\partial t} \right\| dA = \int_{t=0}^{20} \int_{s=0}^{10} \sqrt{14} \, ds \, dt = 200\sqrt{14}.$$

Problems

13. The elliptic cylindrical surface is parameterized by

$$\vec{r} = x\vec{i} + y\vec{j} + z\vec{k} = a\cos\theta\vec{i} + b\sin\theta\vec{j} + z\vec{k} \qquad \text{where } 0 \leq \theta \leq 2\pi, -c \leq z \leq c.$$

We have

$$\frac{\partial \vec{r}}{\partial \theta} \times \frac{\partial \vec{r}}{\partial z} = \begin{vmatrix} \vec{i} & \vec{j} & \vec{k} \\ -a\sin\theta & b\cos\theta & 0 \\ 0 & 0 & 1 \end{vmatrix} = b\cos\theta\vec{i} + a\sin\theta\vec{j}.$$

This vector points away from the z-axis, so we use $d\vec{A} = (b\cos\theta\vec{i} + a\sin\theta\vec{j})\, d\theta dz$, giving

$$\int_S \vec{F} \cdot d\vec{A} = \int_{-c}^{c} \int_0^{2\pi} (\frac{b}{a}(a\cos\theta)\vec{i} + \frac{a}{b}(b\sin\theta\vec{j})) \cdot (b\cos\theta\vec{i} + a\sin\theta\vec{j})\, d\theta \, dz$$

$$= \int_{-c}^{c} \int_0^{2\pi} (b^2\cos^2\theta + a^2\sin^2\theta)\, d\theta dz$$

$$= 2\pi c(a^2 + b^2).$$

17. The surface of S is parameterized by

$$\vec{r}(\theta, \phi) = x\vec{i} + y\vec{j} + z\vec{k},$$

where

$$\begin{cases} x = a + d\sin\phi\cos\theta, \\ y = b + d\sin\phi\sin\theta, \quad \text{for} \quad 0 \leq \phi \leq \pi, 0 \leq \theta \leq 2\pi. \\ z = c + d\cos\phi, \end{cases}$$

The vector $\partial\vec{r}/\partial\phi \times \partial\vec{r}/\partial\theta$ points outward by the right-hand rule, so

$$d\vec{A} = \left(\frac{\partial \vec{r}}{\partial \phi} \times \frac{\partial \vec{r}}{\partial \theta} \right) d\phi d\theta.$$

Thus,

$$\vec{F} \cdot d\vec{A} = \vec{F} \cdot \left(\frac{\partial \vec{r}}{\partial \phi} \times \frac{\partial \vec{r}}{\partial \theta} \right) d\phi d\theta$$

$$= \begin{vmatrix} x^2 & y^2 & z^2 \\ \frac{\partial x}{\partial \phi} & \frac{\partial y}{\partial \phi} & \frac{\partial z}{\partial \phi} \\ \frac{\partial x}{\partial \theta} & \frac{\partial y}{\partial \theta} & \frac{\partial z}{\partial \theta} \end{vmatrix} d\phi d\theta$$

$$= \begin{vmatrix} (a + d\sin\phi\cos\theta)^2 & (b + d\sin\phi\sin\theta)^2 & (c + d\cos\phi)^2 \\ d\cos\phi\cos\theta & d\cos\phi\sin\theta & -d\sin\phi \\ -d\sin\phi\sin\theta & d\sin\phi\cos\theta & 0 \end{vmatrix} d\phi d\theta.$$

Hence,

$$\int_S \vec{F} \cdot d\vec{A} = d^2 \int_0^{2\pi} \int_0^{\pi} (\ a^2 \sin^2 \phi \cos \theta + 2ad \sin^3 \phi \cos^2 \theta + d^2 \sin^4 \phi \cos^3 \theta$$
$$+\ b^2 \sin^2 \phi \sin \theta + 2bd \sin^3 \phi \sin^2 \theta + d^2 \sin^4 \phi \sin^3 \theta$$
$$+\ c^2 \sin \phi \cos \phi + 2cd \sin \phi \cos^2 \phi + d^2 \sin \phi \cos^3 \phi)\ d\phi d\theta.$$

Since

$$\int_0^{2\pi} \cos \theta\ d\theta = \int_0^{2\pi} \sin \theta\ d\theta = \int_0^{2\pi} \cos^3 \theta\ d\theta = \int_0^{2\pi} \sin^3 \theta\ d\theta = 0,$$

and

$$\int_0^{\pi} \sin \phi \cos \phi\ d\phi = \int_0^{\pi} \sin \phi \cos^3 \phi\ d\phi = 0,$$

we have

$$\int_S \vec{F} \cdot \vec{A} \quad d^2 \int_0^{2\pi} \int_0^{\pi} (2ad \sin^3 \phi \cos^2 \theta + 2bd \sin^3 \phi \sin^2 \theta + 2cd \sin \phi \cos^2 \phi)\ d\phi d\theta$$

$$= 2\pi d^3 \int_0^{\pi} (a \sin^3 \phi + b \sin^3 \phi + 2c \sin \phi \cos^2 \phi)\ d\phi$$

$$= 4\pi d^3 \int_0^{\pi/2} (a \sin^3 \phi + b \sin^3 \phi + 2c \sin \phi \cos^2 \phi)\ d\phi$$

$$= \frac{8}{3} \pi d^3 (a + b + c).$$

21. In terms of the st-parameterization,

$$d\vec{A} = \frac{\partial \vec{r}}{\partial s} \times \frac{\partial \vec{r}}{\partial t}\ ds\, dt.$$

By the chain rule, we have

$$\frac{\partial \vec{r}}{\partial s} = \frac{\partial \vec{r}}{\partial u} \frac{\partial u}{\partial s} + \frac{\partial \vec{r}}{\partial v} \frac{\partial v}{\partial s}$$
$$\frac{\partial \vec{r}}{\partial t} = \frac{\partial \vec{r}}{\partial u} \frac{\partial u}{\partial t} + \frac{\partial \vec{r}}{\partial v} \frac{\partial v}{\partial t}.$$

So taking the cross product, we get

$$\frac{\partial \vec{r}}{\partial s} \times \frac{\partial \vec{r}}{\partial t} = \left(\frac{\partial \vec{r}}{\partial u} \frac{\partial u}{\partial s} + \frac{\partial \vec{r}}{\partial v} \frac{\partial v}{\partial s} \right) \times \left(\frac{\partial \vec{r}}{\partial u} \frac{\partial u}{\partial t} + \frac{\partial \vec{r}}{\partial v} \frac{\partial v}{\partial t} \right)$$

$$= \left(\frac{\partial u}{\partial s} \frac{\partial v}{\partial t} - \frac{\partial u}{\partial t} \frac{\partial v}{\partial s} \right) \frac{\partial \vec{r}}{\partial u} \times \frac{\partial \vec{r}}{\partial v}.$$

Now suppose we are going to change variables in a double integral from uv-coordinates to st-coordinates. The Jacobian is

$$\frac{\partial(u, v)}{\partial(s, t)} = \begin{vmatrix} \frac{\partial u}{\partial s} & \frac{\partial v}{\partial s} \\ \frac{\partial u}{\partial t} & \frac{\partial v}{\partial t} \end{vmatrix} = \frac{\partial u}{\partial s} \frac{\partial v}{\partial t} - \frac{\partial u}{\partial t} \frac{\partial v}{\partial s}.$$

Since the Jacobian is assumed to be positive, converting from a uv-integral to an st-integral gives:

$$\int_T \vec{F} \cdot \frac{\partial \vec{r}}{\partial u} \times \frac{\partial \vec{r}}{\partial v}\ du dv = \int_R \vec{F} \cdot \frac{\partial \vec{r}}{\partial u} \times \frac{\partial \vec{r}}{\partial v} \frac{\partial(u, v)}{\partial(s, t)}\ ds dt$$

$$= \int_R \vec{F} \cdot \frac{\partial \vec{r}}{\partial u} \times \frac{\partial \vec{r}}{\partial v} \left(\frac{\partial u}{\partial s} \frac{\partial v}{\partial t} - \frac{\partial u}{\partial t} \frac{\partial v}{\partial s} \right)\ ds dt.$$

However, we know that this gives us

$$\int_T \vec{F} \cdot \frac{\partial \vec{r}}{\partial u} \times \frac{\partial \vec{r}}{\partial v}\ du\, dv = \int_R \vec{F} \cdot \frac{\partial \vec{r}}{\partial u} \times \frac{\partial \vec{r}}{\partial v} \left(\frac{\partial u}{\partial s} \frac{\partial v}{\partial t} - \frac{\partial u}{\partial t} \frac{\partial v}{\partial s} \right)\ ds dt = \int_R \vec{F} \cdot \frac{\partial \vec{r}}{\partial s} \times \frac{\partial \vec{r}}{\partial t}\ ds dt.$$

Thus, the flux integral in uv-coordinates equals the flux integral in st-coordinates.

Solutions for Chapter 19 Review

Exercises

1. Scalar. Since the surface is closed and the vector field is constant, the flux in one side equals the flux out on the other side, so the net flux through the surface is 0.

5. The only contribution to the flux is from the face $z = 3$, since the vector field is zero or parallel to the other faces. On this face, $\vec{H} = 3x\vec{k}$. The vector field is everywhere perpendicular to the face $z = 3$ but varies in magnitude from point to point. On this surface, $d\vec{A} = \vec{k}\, dx\, dy$. Thus

$$\text{Flux} = \int_0^2 \int_0^1 3x\vec{k} \cdot \vec{k}\, dx\, dy = \int_0^2 \int_0^1 3x\, dx\, dy = \left.\frac{3x^2}{2}\right|_0^1 \cdot \left. y\right|_0^2 = 3.$$

9. Only the $\vec{i}$-component contributes to the flux, so

$$\text{Flux} = 2\pi \cdot \text{ Area of surface} = 2\pi(\pi 3^2) = 18\pi^2.$$

13. Since the vector field is everywhere perpendicular to the surface of the sphere, and $||\vec{F}|| = \pi$ on the surface, we have

$$\int_S \vec{F} \cdot d\vec{A} = ||\vec{F}|| \cdot \text{ Area of sphere} = \pi \cdot 4\pi(\pi)^2 = 4\pi^4.$$

17. The area vector of the surface is $-49\vec{j}$, so

$$\int_S \vec{G} \cdot d\vec{A} = (2\vec{j} + 3\vec{k}) \cdot (-49\vec{j}) = -98.$$

21. Only the $\vec{k}$ component contributes to the flux. In the plane $z = 4$, we have $\vec{F} = 2\vec{i} + 3\vec{j} + 4\vec{k}$. On the square $d\vec{A} = \vec{k}\, dA$, so we have

$$\text{Flux} = \int \vec{F} \cdot d\vec{A} = 4\vec{k} \cdot (\vec{k}\ \text{Area of square}) = 4(5^2) = 100.$$

25. See Figure 19.2. The vector field is a vortex going around the z-axis, and the square is centered on the x-axis, so the flux going across one half of the square is balanced by the flux coming back across the other half. Thus, the net flux is zero, so

$$\int_S \vec{F} \cdot d\vec{A} = 0.$$

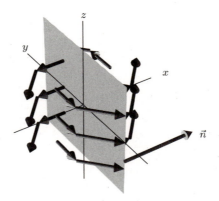

Figure 19.2

29. On the sphere of radius 2, the vector field has $\|\vec{F}\| = 10$ and points inward everywhere (opposite to the orientation of the surface). So

$$\text{Flux} = \int_S \vec{F} \cdot d\vec{A} = -\|\vec{F}\| \cdot \text{Area of sphere} = -10 \cdot 4\pi 2^2 = -160\pi.$$

33. On the surface S, y is constant, $y = -1$, and $d\vec{A} = -\vec{j}\, dA$, so,

$$\int_S \vec{F} \cdot d\vec{A} = \int_S (x^2\vec{i} + (x + e^{-1})\vec{j} - \vec{k}) \cdot (-\vec{j})\, dA = -\int_S (x + e^{-1})\, dA$$

$$= -\int_0^4 \int_0^2 (x + e^{-1})\, dx\, dz = -4(2 + 2e^{-1}) = -8(1 + e^{-1}).$$

37. First we have

$$z_x = \frac{x}{\sqrt{x^2 + y^2}} \qquad z_y = \frac{y}{\sqrt{x^2 + y^2}}.$$

Although z is not a smooth function of x and y at $(0,0)$, the improper integral that we get converges:

$$\int_S \vec{F} \cdot d\vec{A} = \int_S (x^2\vec{i} + y^2\vec{j} + \sqrt{x^2 + y^2}\,\vec{k}) \cdot (-\frac{x}{\sqrt{x^2 + y^2}}\vec{i} - \frac{y}{\sqrt{x^2 + y^2}}\vec{j} + \vec{k})\, dA$$

$$= \int_S \left(-\frac{x^3 + y^3}{\sqrt{x^2 + y^2}} + \sqrt{x^2 + y^2}\right) dA$$

Changing to polar coordinates we have

$$\int_S \vec{F} \cdot d\vec{A} = \int_0^{\pi/2} \int_0^1 (-r^2\cos^3\theta - r^2\sin^3\theta + r)r\, dr d\theta$$

$$= \int_0^{\pi/2} \left(-\frac{r^4}{4}(\cos^3\theta + \sin^3\theta) + \frac{1}{3}r^3 \Big|_{r=0}^{r=1}\right) d\theta$$

$$= \int_0^{\pi/2} \left(-\frac{1}{4}(\cos^3\theta + \sin^3\theta) + \frac{1}{3}\right) d\theta$$

$$= \int_0^{\pi/2} \left(-\frac{1}{4}(\cos\theta - \cos\theta\sin^2\theta + \sin\theta - \sin\theta\cos^2\theta) + \frac{1}{3}\right) d\theta$$

$$= -\frac{1}{4}(\sin\theta - \frac{1}{3}\sin^3\theta - \cos\theta + \frac{1}{3}\cos^3\theta) + \frac{\theta}{3}\Big|_0^{\pi/2}$$

$$= \frac{\pi}{6} - \frac{1}{3}.$$

Problems

41. (a) We have $\operatorname{grad} f = (y + yze^{xyz})\vec{i} + (x + xze^{xyz})\vec{j} + xye^{xyz}\vec{k}$.
   (b) By the Fundamental Theorem of Line Integrals, we have

$$\int_C \operatorname{grad} f \cdot d\vec{r} = (xy + e^{xyz})\Big|_{(1,1,1)}^{(2,3,4)} = (2 \cdot 3 + e^{2 \cdot 3 \cdot 4}) - (1 \cdot 1 + e^{1 \cdot 1 \cdot 1}) = 5 + e^{24} - e^1.$$

   (c) Only the k-component of $\operatorname{grad} f$ contributes to the flux integral. On the xy-plane, $d\vec{A} = \vec{k}\, dx\, dy$ and $z = 0$, so

$$\int_S \operatorname{grad} f \cdot d\vec{A} = \int_0^2 \int_0^{\sqrt{4-x^2}} xye^0\, dy\, dx = \int_0^2 \frac{xy^2}{2}\Big|_0^{\sqrt{4-x^2}} = \int_0^2 \frac{x}{2}(4 - x^2)\, dx = \left(x^2 - \frac{x^4}{8}\right)\Big|_0^2 = 2.$$

45. The vector field $\vec{D}$ has constant magnitude on S, equal to $Q/4\pi R^2$, and points radially outward, so

$$\int_S \vec{D} \cdot d\vec{A} = \frac{Q}{4\pi R^2} \cdot 4\pi R^2 = Q.$$

49. (a) Consider two opposite faces of the cube, S_1 and S_2. The corresponding area vectors are $\vec{A}_1 = 4\vec{i}$ and $\vec{A}_2 = -4\vec{i}$ (since the side of the cube has length 2). Since $\vec{E}$ is constant, we find the flux by taking the dot product, giving

$$\text{Flux through } S_1 = \vec{E} \cdot \vec{A}_1 = (a\vec{i} + b\vec{j} + c\vec{k}) \cdot 4\vec{i} = 4a.$$

$$\text{Flux through } S_2 = \vec{E} \cdot \vec{A}_2 = (a\vec{i} + b\vec{j} + c\vec{k}) \cdot (-4\vec{i}) = -4a.$$

Thus the fluxes through S_1 and S_2 cancel. Arguing similarly, we conclude that, for any pair of opposite faces, the sum of the fluxes of $\vec{E}$ through these faces is zero. Hence, by addition, $\int_S \vec{E} \cdot d\vec{A} = 0$.

(b) The basic idea is the same as in part (a), except that we now need to use Riemann sums. First divide S into two hemispheres H_1 and H_2 by the equator C located in a plane perpendicular to $\vec{E}$. For a tiny patch S_1 in the hemisphere H_1, consider the patch S_2 in the opposite hemisphere which is symmetric to S_1 with respect to the center O of the sphere. The area vectors $\Delta \vec{A}_1$ and $\Delta \vec{A}_2$ satisfy $\Delta \vec{A}_2 = -\Delta \vec{A}_1$, so if we consider S_1 and S_2 to be approximately flat, then $\vec{E} \cdot \Delta \vec{A}_1 = -\vec{E} \cdot \Delta \vec{A}_2$. By decomposing H_1 and H_2 into small patches as above and using Riemann sums, we get

$$\int_{H_1} \vec{E} \cdot d\vec{A} = -\int_{H_2} \vec{E} \cdot d\vec{A}, \quad \text{so} \quad \int_S \vec{E} \cdot d\vec{A} = 0.$$

(c) The reasoning in part (b) can be used to prove that the flux of $\vec{E}$ through any surface with a center of symmetry is zero. For instance, in the case of the cylinder, cut it in half with a plane $z = 1$ and denote the two halves by H_1 and H_2. Just as before, take patches in H_1 and H_2 with $\Delta A_1 = -\Delta A_2$, so that $\vec{E} \cdot \Delta A_1 = -\vec{E} \cdot \Delta \vec{A}_2$. Thus, we get

$$\int_{H_1} \vec{E} \cdot d\vec{A} = -\int_{H_2} \vec{E} \cdot d\vec{A},$$

which shows that

$$\int_S \vec{E} \cdot d\vec{A} = 0.$$

CAS Challenge Problems

53. (a) When $x > 0$, the vector $x\vec{i}$ points in the positive x-direction, and when $x < 0$ it points in the negative x-direction. Thus it always points from the inside of the ellipsoid to the outside, so we expect the flux integral to be positive. The upper half of the ellipsoid is the graph of $z = f(x,y) = \frac{1}{\sqrt{2}}(1 - x^2 - y^2)$, so the flux integral is

$$\int_S \vec{F} \cdot d\vec{A} = \int_{-1/2}^{1/2} \int_{-1/2}^{1/2} x\vec{i} \cdot (-f_x\vec{i} - f_y\vec{j} + \vec{k}) \, dxdy$$

$$= \int_{-1/2}^{1/2} \int_{-1/2}^{1/2} (-xf_x) \, dxdy = \int_{-1/2}^{1/2} \int_{-1/2}^{1/2} \frac{x^2}{\sqrt{1 - x^2 - y^2}} \, dxdy$$

$$= \frac{-\sqrt{2} + 11 \arcsin(\frac{1}{\sqrt{3}}) + 10 \arctan(\frac{1}{\sqrt{2}}) - 8 \arctan(\frac{5}{\sqrt{2}})}{12} = 0.0958.$$

Different CASs may give the answer in different forms. Note that we could have predicted the integral was positive without evaluating it, since the integrand is positive everywhere in the region of integration.

(b) For $x > -1$, the quantity $x + 1$ is positive, so the vector field $(x + 1)\vec{i}$ always points in the direction of the positive x-axis. It is pointing into the ellipsoid when $x < 0$ and out of it when $x > 0$. However, its magnitude is smaller when $-1/2 < x < 0$ than it is when $0 < x < 1/2$, so the net flux out of the ellipsoid should be positive. The flux integral is

$$\int_S \vec{F} \cdot d\vec{A} = \int_{-1/2}^{1/2} \int_{-1/2}^{1/2} (x + 1)\vec{i} \cdot (-f_x\vec{i} - f_y\vec{j} + \vec{k}) \, dxdy$$

$$= \int_{-1/2}^{1/2} \int_{-1/2}^{1/2} -(x + 1)f_x \, dxdy = \int_{-1/2}^{1/2} \int_{-1/2}^{1/2} \frac{x(1 + x)}{\sqrt{1 - x^2 - y^2}} \, dxdy$$

$$= \frac{\sqrt{2} - 11\,\arcsin(\frac{1}{\sqrt{3}}) - 10\,\arctan(\frac{1}{\sqrt{2}}) + 8\,\arctan(\frac{5}{\sqrt{2}})}{12} = 0.0958$$

The answer is the same as in part (a). This makes sense because the difference between the integrals in parts (a) and (b) is the integral of $\int_{-1/2}^{1/2}\int_{-1/2}^{1/2}(x/\sqrt{1-x^2-y^2})\,dx\,dy$, which is zero because the integrand is odd with respect to x.

(c) This integral should be positive for the same reason as in part (a). The vector field $y\vec{j}$ points in the positive y-direction when $y > 0$ and in the negative y-direction when $y < 0$, thus it always points out of the ellipsoid. Evaluating the integral we get

$$\int_S \vec{F}\cdot d\vec{A} = \int_{-1/2}^{1/2}\int_{-1/2}^{1/2} y\vec{j}\cdot(-f_x\vec{i} - f_y\vec{j} + \vec{k})\,dx\,dy$$

$$= \int_{-1/2}^{1/2}\int_{-1/2}^{1/2}(-yf_y)\,dx\,dy = \int_{-1/2}^{1/2}\int_{-1/2}^{1/2}\frac{y^2}{\sqrt{1-x^2-y^2}}\,dx\,dy$$

$$= \frac{\sqrt{2} - 2\,\arcsin(\frac{1}{\sqrt{3}}) - 19\,\arctan(\frac{1}{\sqrt{2}}) + 8\,\arctan(\frac{5}{\sqrt{2}})}{12} = 0.0958.$$

The symbolic answer appears different but has the same numerical value as in parts (a) and (b). In fact the answer is the same because the integral here is the same as in part (a) except that the roles of x and y have been exchanged. Different CASs may give different symbolic forms.

CHECK YOUR UNDERSTANDING

1. True. By definition, the flux integral is the limit of a sum of dot products, hence is a scalar.

5. True. The flow of this field is in the same direction as the orientation of the surface everywhere on the surface, so the flux is positive.

9. True. In the sum defining the flux integral for $\vec{F}$, we have terms like $\vec{F}\cdot\Delta\vec{A} = (2\vec{G})\cdot\Delta\vec{A} = 2(\vec{G}\cdot\Delta\vec{A})$. So each term in the sum approximating the flux of $\vec{F}$ is twice the corresponding term in the sum approximating the flux of $\vec{G}$, making the sum for $\vec{F}$ twice that of the sum for $\vec{G}$. Thus the flux of $\vec{F}$ is twice the flux of $\vec{G}$.

13. False. Both surfaces are oriented upward, so $\vec{A}(x, y)$ and $\vec{B}(x, y)$ both point upward. But they could point in different directions, since the graph of $z = -f(x, y)$ is the graph of $z = f(x, y)$ turned upside down.

CHAPTER TWENTY

Solutions for Section 20.1

Exercises

1. Scalar. Since

$$\operatorname{div}\left(\frac{y\vec{i} - x\vec{j}}{x^2 + y^2}\right) = \frac{-y \cdot 2x}{(x^2 + y^2)^2} + \frac{x \cdot 2y}{(x^2 + y^2)^2} = 0.$$

5. $\operatorname{div} \vec{F} = \dfrac{\partial}{\partial x}(-x + y) + \dfrac{\partial}{\partial y}(y + z) + \dfrac{\partial}{\partial z}(-z + x) = -1 + 1 - 1 = -1$

9. Using the formula for $\vec{a} \times \vec{r}$ in Cartesian coordinates, we get

$$\operatorname{div} \vec{F} = \frac{\partial}{\partial x}(a_2 z - a_3 y) + \frac{\partial}{\partial y}(a_3 x - a_1 z) + \frac{\partial}{\partial z}(a_1 y - a_2 x) = 0$$

13. Two vector fields that have positive divergence everywhere are in Figures 20.1 and 20.2.

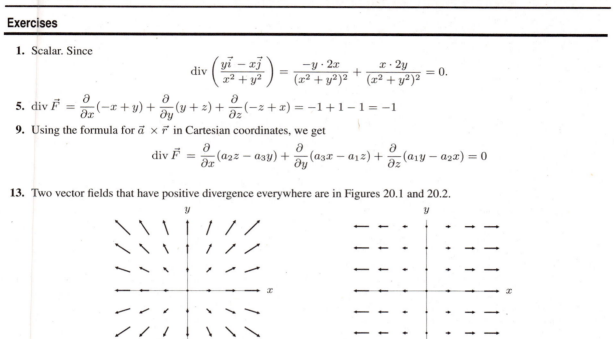

Figure 20.1   Figure 20.2

Problems

17. Since div $F(1, 2, 3)$ is the flux density out of a small region surrounding the point $(1, 2, 3)$, we have

$$\operatorname{div} \vec{F}(1, 2, 3) \approx \frac{\text{Flux out of small region around } (1, 2, 3)}{\text{Volume of region.}}$$

So

$$\text{Flux out of region} \approx (\operatorname{div} \vec{F}(1, 2, 3)) \cdot \text{Volume of region}$$
$$= 5 \cdot \frac{4}{3}\pi(0.01)^3$$
$$= \frac{0.00002\pi}{3}.$$

21. Using flux: On S_1, $x = a$ and normal is in negative x-direction, so

$$\vec{F} \cdot \Delta \vec{A} = ((3a + 2)\vec{i} + 4a\vec{j} + (5a + 1)\vec{k}) \cdot (-\Delta A \vec{i}) = -(3a + 2)\Delta A$$

Thus

$$\int_{S_1} \vec{F} \cdot d\vec{A} = \int_{S_1} -(3a + 2)dA = -(3a + 2)(\text{Area of } S_1) = -(3a + 2)w^2.$$

On S_2, $x = a + w$ and normal is in the positive x-direction, so

$$\vec{F} \cdot \Delta \vec{A} = [(3(a + w) + 2)\vec{i} + 4(a + w)\vec{j} + (5(a + w) + 1)\vec{k}] \cdot (\Delta A \vec{i}) = (3a + 3w + 2)\Delta A.$$

Thus

$$\int_{S_2} \vec{F} \cdot d\vec{A} = \int_{S_2} (3a + 3w + 2)dA = (3a + 3w + 2)(\text{Area of } S_2) = (3a + 3w + 2)w^2.$$

Next, we have $\int_{S_3} \vec{F} \cdot d\vec{A} = \int_{S_3} -4x\,dA$ and $\int_{S_4} \vec{F} \cdot d\vec{A} = \int_{S_4} 4x\,dA$. Since these two are integrated over the same region in the xz-plane, the two integrals cancel. Similarly, $\int_{S_5} \vec{F} \cdot d\vec{A} = \int_{S_5} -(5x+1)\,d\vec{A}$ cancels out $\int_{S_6} \vec{F} \cdot d\vec{A} = \int_{S_6} (5x+1)\,d\vec{A}$. Therefore,

Total flux

$$= \int_{S_1} \vec{F} \cdot d\vec{A} + \int_{S_2} \vec{F} \cdot d\vec{A} + \int_{S_3} \vec{F} \cdot d\vec{A} + \int_{S_4} \vec{F} \cdot d\vec{A} + \int_{S_5} \vec{F} \cdot d\vec{A} + \int_{S_6} \vec{F} \cdot d\vec{A}$$

$$= -(3a+2)w^2 + (3a+3w+2)w^2 + \int_{S_3} -4x\,dA + \int_{S_4} 4x\,dA$$

$$+ \int_{S_5} -(5x+1)\,dA + \int_{S_6} (5x+1)\,dA = 3w^3.$$

To find div $\vec{F}$ at the point (a, b, c), let the box shrink to the point by letting $w \to 0$. Then

$$\text{div } \vec{F} = \lim_{w \to 0} \left(\frac{\text{Flux through box}}{\text{Volume of box}} \right)$$

$$= \lim_{w \to 0} \left(\frac{3w^3}{w^3} \right) = 3.$$

Using partial derivatives:

$$\text{div } \vec{F} = \frac{\partial}{\partial x}(3x+2) + \frac{\partial}{\partial y}(4x) + \frac{\partial}{\partial z}(5x+1) = 3$$

25. (a) div $\vec{B} = \dfrac{\partial}{\partial x}(-y) + \dfrac{\partial}{\partial y}(x) + \dfrac{\partial}{\partial z}(x+y) = 0$, so this could be a magnetic field.

(b) div $\vec{B} = \dfrac{\partial}{\partial x}(-z) + \dfrac{\partial}{\partial y}(y) + \dfrac{\partial}{\partial z}(x) = 0 + 1 + 0 = 1$, so this could not be a magnetic field.

(c) div $\vec{B} = \dfrac{\partial}{\partial x}(x^2 - y^2 - x) + \dfrac{\partial}{\partial y}(y - 2xy) + \dfrac{\partial}{\partial z}(0) = 2x - 1 + 1 - 2x + 0 = 0$, so this could be a magnetic field.

29. Let $\vec{F} = F_1\vec{i} + F_2\vec{j} + F_3\vec{k}$. Then

$$\text{div}(g\vec{F}) = \text{div}(gF_1\vec{i} + gF_2\vec{j} + gF_3\vec{k})$$

$$= \frac{\partial}{\partial x}(gF_1) + \frac{\partial}{\partial y}(gF_2) + \frac{\partial}{\partial z}(gF_3)$$

$$= \frac{\partial g}{\partial x}F_1 + g\frac{\partial F_1}{\partial x} + \frac{\partial g}{\partial y}F_2 + g\frac{\partial F_2}{\partial y} + \frac{\partial g}{\partial z}F_3 + g\frac{\partial F_3}{\partial z}$$

$$= \frac{\partial g}{\partial x}F_1 + \frac{\partial g}{\partial y}F_2 + \frac{\partial g}{\partial z}F_3 + g\left(\frac{\partial F_1}{\partial x} + \frac{\partial F_2}{\partial y} + \frac{\partial F_3}{\partial z}\right)$$

$$= (\text{grad } g) \cdot \vec{F} + g\,\text{div }\vec{F}.$$

33. Using $\text{div}(g\vec{F}) = (\text{grad } g) \cdot \vec{F} + g\,\text{div }\vec{F}$, we have

$$\text{div }\vec{G} = \text{grad}(\vec{b} \cdot \vec{r}) \cdot (\vec{a} \times \vec{r}) + \vec{b} \cdot \vec{r}\,\text{div}(\vec{a} \times \vec{r}) = \vec{b} \cdot (\vec{a} \times \vec{r}) + \vec{v} \cdot \vec{r}\,0 = \vec{b} \cdot (\vec{a} \times \vec{r}).$$

37. (a) The velocity vector for the traffic flow would look like:

(b) When $0 \le x < 2000$, the velocity is decreasing linearly from 55 to 15, so its formula is $(55 - x/50)\vec{i}$ mph. Then, when $2000 \le x < 7000$, the speed is constant, so $\vec{v}(x) = 15\vec{i}$ mph. Next, when $7000 \le x < 8000$, the velocity is increasing linearly from 15 to 55, so $\vec{v}(x) = (15 + (x - 7000)/25)\vec{i}$ mph. Finally, when $x \ge 8000$, the speed is constant, so $\vec{v}(x) = 55\vec{i}$ mph.

(c) $\operatorname{div} \vec{v} = dv(x)/dx$.

At $x = 1000$, $v(x) = 55 - x/50$, so $\operatorname{div} \vec{v} = -1/50$.

At $x = 5000$, $v(x) = 15$, so $\operatorname{div} \vec{v} = 0$.

At $x = 7500$, $v(x) = 15 + (x - 7000)/25$, so $\operatorname{div} \vec{v} = 1/25$.

At $x = 10,000$, $v(x) = 55$, so $\operatorname{div} \vec{v} = 0$.

In each case the units of $\operatorname{div} \vec{v}$ are $\dfrac{\text{miles/hour}}{\text{feet}}$.

41. (a) At any point $\vec{r} = x\vec{i} + y\vec{j}$, the direction of the vector field $\vec{v}$ is pointing toward the origin, which means it is of the form $\vec{v} = f\vec{r}$ for some negative function f whose value can vary depending on $\vec{r}$. The magnitude of $\vec{v}$ depends only on the distance r, thus f must be a function depending only on r, which is equivalent to depending only on r^2 since $r \geq 0$. So $\vec{v} = f(r^2)\vec{r} = \left(f(x^2 + y^2)\right)(x\vec{i} + y\vec{j})$.

(b) At $(x, y) \neq (0, 0)$ the divergence of $\vec{v}$ is

$$\operatorname{div} \vec{v} = \frac{\partial (K(x^2 + y^2)^{-1}x)}{\partial x} + \frac{\partial (K(x^2 + y^2)^{-1}y)}{\partial y} = \frac{Ky^2 - Kx^2}{(x^2 + y^2)^2} + \frac{Kx^2 - Ky^2}{(x^2 + y^2)^2} = 0.$$

Therefore, $\vec{v}$ is a point sink at the origin.

(c) The magnitude of $\vec{v}$ is

$$\|\vec{v}\| = |K|(x^2 + y^2)^{-1}|x\vec{i} + y\vec{j}| = |K|(x^2 + y^2)^{-1}(x^2 + y^2)^{1/2} = |K|(x^2 + y^2)^{-1/2} = \frac{|K|}{r}.$$

(remember, $K < 0$)

(d) The vector field looks like the following:

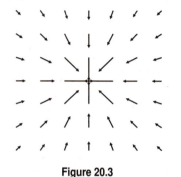

Figure 20.3

(e) We need to show that $\operatorname{grad} \phi = \vec{v}$.

$$\operatorname{grad} \phi = \frac{\partial}{\partial x}\left(\frac{K}{2}\log(x^2 + y^2)\right)\vec{i} + \frac{\partial}{\partial y}\left(\frac{K}{2}\log(x^2 + y^2)\right)\vec{j}$$

$$= \frac{Kx}{x^2 + y^2}\vec{i} + \frac{Ky}{x^2 + y^2}\vec{j}$$

$$= K(x^2 + y^2)^{-1}(x\vec{i} + y\vec{j})$$

$$= \vec{v}$$

Solutions for Section 20.2

Exercises

1. First directly: On the faces $x = 0$, $y = 0$, $z = 0$, the flux is zero. On the face $x = 2$, a unit normal is $\vec{i}$ and $d\vec{A} = dA\vec{i}$. So

$$\int_{S_{x=2}} \vec{r} \cdot d\vec{A} = \int_{S_{x=2}} (2\vec{i} + y\vec{j} + z\vec{k}) \cdot (dA\vec{i})$$

(since on that face, $x = 2$)

$$= \int_{S_{x=2}} 2 \, dA = 2 \cdot (\text{Area of face}) = 2 \cdot 4 = 8.$$

In exactly the same way, you get

$$\int_{S_{y=2}} \vec{r} \cdot d\vec{A} = \int_{S_{z=2}} \vec{r} \cdot d\vec{A} = 8,$$

so

$$\int_S \vec{r} \cdot d\vec{A} = 3 \cdot 8 = 24.$$

Now using divergence:

$$\operatorname{div} \vec{F} = \frac{\partial x}{\partial x} + \frac{\partial y}{\partial y} + \frac{\partial z}{\partial z} = 3,$$

so

$$\text{Flux} = \int_0^2 \int_0^2 \int_0^2 3 \, dx \, dy \, dz = 3 \cdot (\text{Volume of Cube}) = 3 \cdot 8 = 24$$

5. The location of the pyramid has not been completely specified. For instance, where is it centered on the xy plane? How is base oriented with respect to the axes? Thus, we cannot compute the flux by direct integration with the information we have. However, we can calculate it using the divergence theorem. First we calculate the divergence of $\vec{F}$.

$$\operatorname{div} \vec{F} = \frac{\partial(-z)}{\partial x} + \frac{\partial 0}{\partial y} + \frac{\partial x}{\partial z} = 0 + 0 + 0 = 0$$

Thus for any closed surface the flux will be zero, so the flux through our pyramid, regardless of its location or orientation, is zero.

9. Since $\operatorname{div} \vec{G} = 1$, if W is the interior of the box, the Divergence Theorem gives

$$\text{Flux} = \int_W 1 \, dV = 1 \cdot \text{Volume of box} = 1 \cdot 2 \cdot 3 \cdot 4 = 24.$$

13. We have

$$\operatorname{div}((3x + 4y)\vec{i} + (4y + 5z)\vec{j} + (5z + 3x)\vec{k}) = 3 + 4 + 5 = 12.$$

Let W be the interior of the cube. Then by the divergence theorem,

$$\int_S ((3x + 4y)\vec{i} + (4y + 5z)\vec{j} + (5z + 3x)\vec{k}) \cdot d\vec{A} = \int_W 12 \, dV = 12 \cdot \text{Volume of cube} = 12 \cdot (2 \cdot 3 \cdot 4) = 288.$$

Problems

17. Since $\operatorname{div} \vec{F} = 3x^2 + 3y^2 + 3z^2$, the Divergence Theorem gives

$$\text{Flux} = \int_S \vec{F} \cdot d\vec{A} = \int_W (3x^2 + 3y^2 + 3z^2) \, dV.$$

In spherical coordinates, the region W lies between the spheres $\rho = 2$ and $\rho = 3$ and inside the cone $\phi = \pi/4$. Since $3x^2 + 3y^2 + 3z^2 = 3\rho^2$, we have

$$\text{Flux} = \int_S \vec{F} \cdot d\vec{A} = \int_0^{2\pi} \int_0^{\pi/4} \int_2^3 3\rho^2 \cdot \rho^2 \sin\phi \, d\rho \, d\phi \, d\theta$$

$$= 2\pi \cdot \frac{3}{5}\rho^5 \Big|_2^3 (-\cos\phi) \Big|_0^{\pi/4} = \frac{633(2 - \sqrt{2})}{5} \pi = 232.98.$$

21. By the Divergence Theorem, $\int_S \vec{F} \cdot d\vec{A} = \int_W \operatorname{div} \vec{F} \, dV = \int_W 0 \, dV = 0$ for a closed surface S, where W is the region enclosed by S.

25. We consider a sphere of radius R centered at the origin and compute the flux of $\vec{r}$ through its surface. Since $\vec{r}$ and $d\vec{A}$ both point radially outward, $\vec{r} \cdot d\vec{A} = \|\vec{r}\| \|d\vec{A}\| = R \, dA$ on the surface of the sphere, so

$$\int_S \vec{F} \cdot d\vec{A} = \int_S R \, dA = R \int_S dA = R(\text{Surface area of a sphere}) = R(4\pi R^2) = 4\pi R^3.$$

Therefore, volume of sphere $= \frac{1}{3} \int_S \vec{F} \cdot d\vec{A} = \frac{4}{3}\pi R^3$.

29. (a) Since $\vec{F}$ is radial, it is everywhere parallel to the area vector, $\Delta \vec{A}$. Also, $||\vec{F}|| = 1$ on the surface of the sphere $x^2 + y^2 + z^2 = 1$, so

$$
\text{Flux through the sphere} = \int_S \vec{F} \cdot d\vec{A} = \lim_{||\Delta \vec{A}|| \to 0} \sum \vec{F} \cdot \Delta \vec{A}
$$

$$
= \lim_{||\Delta \vec{A}|| \to 0} \sum ||\vec{F}|| \, ||\Delta \vec{A}|| = \lim_{||\Delta \vec{A}|| \to 0} \sum ||\Delta \vec{A}||
$$

$$
= \text{Surface area of sphere} = 4\pi \cdot 1^2 = 4\pi.
$$

(b) In Cartesian coordinates,

$$
\vec{F}(x, y, z) = \frac{x}{(x^2 + y^2 + z^2)^{3/2}} \vec{i} + \frac{y}{(x^2 + y^2 + z^2)^{3/2}} \vec{j} + \frac{z}{(x^2 + y^2 + z^2)^{3/2}} \vec{k}.
$$

So,

$$
\text{div} \, \vec{F}(x, y, z) = \left(\frac{1}{(x^2 + y^2 + z^2)^{3/2}} - \frac{3x^2}{(x^2 + y^2 + z^2)^{5/2}} \right)
$$

$$
+ \left(\frac{1}{(x^2 + y^2 + z^2)^{3/2}} - \frac{3y^2}{(x^2 + y^2 + z^2)^{5/2}} \right)
$$

$$
+ \left(\frac{1}{(x^2 + y^2 + z^2)^{3/2}} - \frac{3z^2}{(x^2 + y^2 + z^2)^{5/2}} \right)
$$

$$
= \left(\frac{x^2 + y^2 + z^2}{(x^2 + y^2 + z^2)^{5/2}} - \frac{3x^2}{(x^2 + y^2 + z^2)^{5/2}} \right)
$$

$$
+ \left(\frac{x^2 + y^2 + z^2}{(x^2 + y^2 + z^2)^{5/2}} - \frac{3y^2}{(x^2 + y^2 + z^2)^{5/2}} \right)
$$

$$
+ \left(\frac{x^2 + y^2 + z^2}{(x^2 + y^2 + z^2)^{5/2}} - \frac{3z^2}{(x^2 + y^2 + z^2)^{5/2}} \right)
$$

$$
= \frac{3(x^2 + y^2 + z^2) - 3(x^2 + y^2 + z^2)}{(x^2 + y^2 + z^2)^{5/2}}
$$

$$
= 0.
$$

(c) We cannot apply the Divergence Theorem to the whole region within the box, because the vector field $\vec{F}$ is not defined at the origin. However, we can apply the Divergence Theorem to the region, W, between the sphere and the box. Since div $\vec{F} = 0$ there, the theorem tells us that

$$
\int_{\substack{\text{Box} \\ \text{(outward)}}} \vec{F} \cdot d\vec{A} + \int_{\substack{\text{Sphere} \\ \text{(inward)}}} \vec{F} \cdot d\vec{A} = \int_W \text{div} \, \vec{F} \, dV = 0.
$$

Therefore, the flux through the box and the sphere are equal if both are oriented outward:

$$
\int_{\substack{\text{Box} \\ \text{(outward)}}} \vec{F} \cdot d\vec{A} = - \int_{\substack{\text{Sphere} \\ \text{(inward)}}} \vec{F} \cdot d\vec{A} = \int_{\substack{\text{Sphere} \\ \text{(outward)}}} \vec{F} \cdot d\vec{A} = 4\pi.
$$

33. (a) At the point $(1, 2, 1)$, we have div $\vec{F} = 1 \cdot 2 \cdot 1^2 = 2$.

(b) Since the box is small, we use the approximation

$$
\text{div} \, \vec{F} = \text{Flux density} \approx \frac{\text{Flux out of box}}{\text{Volume of box}}.
$$

Thus

$$
\text{Flux out of box} \approx (\text{div} \, \vec{F}) \cdot (\text{Volume of box}) = 2(0.2)^3 = 0.016.
$$

(c) To calculate the flux exactly, we use the Divergence Theorem,

$$
\text{Flux out of box} = \int_{\text{Box}} \text{div} \, \vec{F} \, dV = \int_{\text{Box}} xyz^2 \, dV.
$$

Since the box has side 0.2, it is given by $0.9 < x < 1.1, 1.9 < y < 2.1, 0.9 < z < 1.1$, so

$$
\text{Flux} = \int_{0.9}^{1.1} \int_{1.9}^{2.1} \int_{0.9}^{1.1} xyz^2 \, dz\,dy\,dx = \int_{0.9}^{1.1} \int_{1.9}^{2.1} xy \left. \frac{z^3}{3} \right|_{0.9}^{1.1} dy\,dx
$$

$$
= \frac{(1.1)^3 - (0.9)^3}{3} \int_{0.9}^{1.1} \left. \frac{xy^2}{2} \right|_{1.9}^{2.1} dx = \frac{(1.1)^3 - (0.9)^3}{3} \cdot \frac{(2.1)^2 - (1.9)^2}{2} \cdot \left. \frac{x^2}{2} \right|_{0.9}^{1.1}
$$

$$
= \frac{(1.1)^3 - (0.9)^3}{3} \cdot \frac{(2.1)^2 - (1.9)^2}{2} \cdot \frac{(1.1)^2 - (0.9)^2}{2} = 0.016053\ldots.
$$

Notice that you can calculate the flux without knowing the vector field, $\vec{F}$.

37. (a) The rate at which heat is generated at any point in the earth is $\text{div} \, \vec{F}$ at that point. So $\text{div} \, \vec{F} = 30$ watts/km^3.
  (b) Differentiating gives $\text{div}(\alpha(x\vec{i} + y\vec{j} + z\vec{k})) = \alpha(1 + 1 + 1) = 3\alpha$ so $\alpha = 30/3 = 10$ watts/km^3. Thus, $\vec{F} = \alpha\vec{r}$ has constant divergence. Note that $\vec{F} = \alpha\vec{r}$ has flow lines going radially outward, and symmetric about the origin.
  (c) The vector $\text{grad} \, T$ gives the direction of greatest increase in temperature. Thus, $-\text{grad} \, T$ gives the direction of greatest decrease in temperature. The equation $\vec{F} = -k \, \text{grad} \, T$ says that heat will flow in the direction of greatest decrease in temperature (i.e. from hot regions to cold), and at a rate proportional to the temperature gradient.
  (d) We assume that $\vec{F}$ is given by the answer to part (b). Then, using part (c), we have

$$
\vec{F} = 10(x\vec{i} + y\vec{j} + z\vec{k}) = -30{,}000 \, \text{grad} \, T,
$$

so

$$
\text{grad} \, T = -\frac{10}{30{,}000}(x\vec{i} + y\vec{j} + z\vec{k}).
$$

Integrating we get

$$
T = \frac{-10}{2(30{,}000)}(x^2 + y^2 + z^2) + C.
$$

At the surface of the earth, $x^2 + y^2 + z^2 = 6400^2$, and $T = 20°C$, so

$$
T = \frac{-1}{6000}(6400^2) + C = 20.
$$

Thus,

$$
C = 20 + \frac{6400^2}{6000} = 6847.
$$

At the center of the earth, $x^2 + y^2 + z^2 = 0$, so

$$
T = 6847°C.
$$

41. Check that $\text{div} \, \vec{E} = 0$ by taking partial derivatives. For instance,

$$
\frac{\partial E_1}{\partial x} = \frac{\partial}{\partial x} [q(x - x_0)[(x - x_0)^2 + (y - y_0)^2 + (z - z_0)^2]^{-3/2}]
$$

$$
= q[(y - y_0)^2 + (z - z_0)^2 - 2(x - x_0)^2][(x - x_0)^2 + (y - y_0)^2 + (z - z_0)^2]^{-5/2}
$$

and similarly,

$$
\frac{\partial E_2}{\partial y} = q[(x - x_0)^2 + (z - z_0)^2 - 2(y - y_0)^2][(x - x_0)^2 + (y - y_0)^2 + (z - z_0)^2]^{-5/2}
$$

$$
\frac{\partial E_3}{\partial z} = q[(x - x_0)^2 + (y - y_0)^2 - 2(z - z_0)^2][(x - x_0)^2 + (y - y_0)^2 + (z - z_0)^2]^{-5/2}.
$$

Therefore,

$$
\frac{\partial E_1}{\partial x} + \frac{\partial E_2}{\partial y} + \frac{\partial E_3}{\partial z} = 0.
$$

The vector field $\vec{E}$ is defined everywhere but at the point with position vector $\vec{r}_0$. If this point lies outside the surface S, the Divergence Theorem can be applied to the region R enclosed by S, yielding:

$$
\int_S \vec{E} \cdot d\vec{A} = \int_R \text{div} \, \vec{E} \, dV = 0.
$$

If the charge q is located inside S, consider a small sphere S_a centered at q and contained in R. The Divergence Theorem for the region R' between the two spheres yields:

$$\int_S \vec{E} \cdot d\vec{A} + \int_{S_a} \vec{E} \cdot d\vec{A} = \int_{R'} \text{div } \vec{E} \ dV = 0.$$

In this formula, the Divergence Theorem requires S to be given the outward orientation, and S_a the inward orientation. To compute $\int_{S_a} \vec{E} \cdot d\vec{A}$, we use the fact that on the surface of the sphere, $\vec{E}$ and $\Delta\vec{A}$ are parallel and in opposite directions, so

$$\vec{E} \cdot \Delta\vec{A} = -\|\vec{E}\| \|\Delta\vec{A}\|$$

since on the surface of a sphere of radius a,

$$\|\vec{E}\| = q\frac{\|\vec{r} - \vec{r}_0\|}{\|\vec{r} - \vec{r}_0\|^3} = \frac{q}{a^2}.$$

Then,

$$\int_{S_a} \vec{E} \cdot d\vec{A} = \int -\frac{q}{a^2} \|d\vec{A}\| = \frac{-q}{a^2} \cdot \text{Surface area of sphere} = -\frac{q}{a^2} \cdot 4\pi a^2 = -4\pi q.$$

$$\int_{S_a} \vec{E} \cdot d\vec{A} = -4\pi q.$$

$$\int_S \vec{E} \cdot d\vec{A} - \int_{S_a} \vec{E} \cdot d\vec{A} = 4\pi q.$$

Solutions for Section 20.3

Exercises

1. Vector. We have

$$\text{curl}(z\vec{i} - x\vec{j} + y\vec{k}) = \begin{vmatrix} \vec{i} & \vec{j} & \vec{k} \\ \frac{\partial}{\partial x} & \frac{\partial}{\partial y} & \frac{\partial}{\partial z} \\ z & -x & y \end{vmatrix} = \vec{i} + \vec{j} - \vec{k}.$$

5. Using the definition of Cartesian coordinates,

$$\text{curl } \vec{F} = \begin{vmatrix} \vec{i} & \vec{j} & \vec{k} \\ \frac{\partial}{\partial x} & \frac{\partial}{\partial y} & \frac{\partial}{\partial z} \\ (-x+y) & (y+z) & (-z+x) \end{vmatrix}$$

$$= \left(\frac{\partial}{\partial y}(-z+x) - \frac{\partial}{\partial z}(y+z)\right)\vec{i} + \left(-\frac{\partial}{\partial x}(-z+x) + \frac{\partial}{\partial z}(-x+y)\right)\vec{j}$$

$$+ \left(\frac{\partial}{\partial x}(y+z) - \frac{\partial}{\partial y}(-x+y)\right)\vec{k}$$

$$= -\vec{i} - \vec{j} - \vec{k}.$$

9. This vector field points radically outward and has unit length everywhere (except the origin). Thus, we would expect its curl to be $\vec{0}$. Computing the curl directly we get

$$\text{curl}\left(\frac{\vec{r}}{\|\vec{r}\|}\right) = \begin{vmatrix} \vec{i} & \vec{j} & \vec{k} \\ \frac{\partial}{\partial x} & \frac{\partial}{\partial y} & \frac{\partial}{\partial z} \\ \frac{x}{(x^2+y^2+z^2)^{1/2}} & \frac{y}{(x^2+y^2+z^2)^{1/2}} & \frac{z}{(x^2+y^2+z^2)^{1/2}} \end{vmatrix}$$

The $\vec{i}$-component is given by $= \left(-\frac{1}{2} \cdot \frac{2yz}{(x^2+y^2+z^2)^{3/2}} - \left(-\frac{1}{2} \cdot \frac{2yz}{(x^2+y^2+z^2)^{1/2}}\right)\right)\vec{i}$

$$= \vec{0}$$

Similarly, the $\vec{j}$ and $\vec{k}$ components are also both $\vec{0}$.

13. This vector field shows no rotation, and the circulation around any closed curve appears to be zero, so the vector field has zero curl.

Problems

17. Yes. $\vec{F} = (1 + y^2)\vec{i}$ has constant direction and curl $\vec{F} = -2y\vec{k} \neq \vec{0}$. The flow lines of a vector field do not have to bend for it to have nonzero curl.

21. The curl is defined in such a way that if $\vec{n}$ is a unit vector and C is a small circle in the plane perpendicular to $\vec{n}$ and with orientation induced by $\vec{n}$, then

$$(\text{curl}\,\vec{G}) \cdot \vec{n} = \text{Circulation density}$$

$$\approx \frac{\int_C \vec{G} \cdot d\vec{r}}{\text{Area inside } C}$$

so

$$\text{Circulation} = \int_C \vec{G} \cdot d\vec{r} \approx \left((\text{curl}\,\vec{G}) \cdot \vec{n}\right) \cdot \text{Area inside } C.$$

(a) Let C be the circle in the xy-plane, and let $\vec{n} = \vec{k}$. Then

$$\text{Circulation} \approx (2\vec{i} - 3\vec{j} + 5\vec{k}) \cdot \vec{k} \cdot \pi(0.01)^2$$

$$= 0.0005\pi.$$

(b) By a similar argument to part (a), with $\vec{n} = \vec{i}$, we find the circulation around the circle in the yz-plane:

$$\text{Circulation} \approx (2\vec{i} - 3\vec{j} + 5\vec{k}) \cdot \vec{i} \cdot \pi(0.01)^2$$

$$= 0.0002\pi.$$

(c) Similarly for circulations around the circle in the xz-plane,

$$\text{Circulation} \approx -0.0003\pi.$$

25. Let $\vec{C} = a\vec{i} + b\vec{j} + c\vec{k}$. Then

$$\text{curl}(\vec{F} + \vec{C}) = \left(\frac{\partial}{\partial y}(F_3 + c) - \frac{\partial}{\partial z}(F_2 + b)\right)\vec{i} + \left(\frac{\partial}{\partial z}(F_1 + a) - \frac{\partial}{\partial x}(F_3 + c)\right)\vec{j}$$

$$+ \left(\frac{\partial}{\partial x}(F_2 + b) - \frac{\partial}{\partial y}(F_1 + a)\right)\vec{k}$$

$$= \left(\frac{\partial F_3}{\partial y} - \frac{\partial F_2}{\partial z}\right)\vec{i} + \left(\frac{\partial F_1}{\partial z} - \frac{\partial F_3}{\partial x}\right)\vec{j} + \left(\frac{\partial F_2}{\partial x} - \frac{\partial F_1}{\partial y}\right)\vec{k}$$

$$= \text{curl}\,\vec{F}.$$

29. By Problem 28, curl $\vec{F} = \text{grad}\,f \times \text{grad}\,g + f\,\text{curl}\,\text{grad}\,g = \text{grad}\,f \times \text{grad}\,g$, since curl grad $g = 0$. Since the cross product of two vectors is perpendicular to both vectors, curl $\vec{F}$ is perpendicular to grad g. But $\vec{F}$ is a scalar times grad g, so curl $\vec{F}$ is perpendicular to $\vec{F}$.

33. Investigate the velocity vector field of the atmosphere near the fire. If the curl of this vector field is non-zero, there is circulatory motion. Consequently, if the magnitude of the curl of this vector field is large near the fire, a fire storm has probably developed.

37. (a) Since $\vec{F}$ is in the xy-plane, curl $\vec{F}$ is parallel to $\vec{k}$ (because $F_3 = 0$ and F_1, F_2 have no z-dependence). Imagine computing the circulation of $\vec{F}$ counterclockwise around a small rectangle R at the point P with sides of length h parallel to $\vec{F}$ and sides of length t perpendicular to $\vec{F}$ as shown in Figure 20.4. Since $\vec{F}$ is perpendicular to C_2 and C_4, the line integral over these two sides is zero. Assuming that $\vec{F}$ is approximately constant on C_1 and C_3, its value on these sides is $F(Q)\vec{T}$ and $-F(P)\vec{T}$, respectively. Thus, since $\vec{F}$ is parallel to C_1 and C_3, the line integral over C_1 is approximately $F(Q)h$ and the line integral over C_3 is approximately $-F(P)h$. Finally

$$\text{curl}\,\vec{F}(P) \approx \frac{\text{Circulation around } R}{\text{Area of } R} \approx \frac{F(Q)h - F(P)h}{ht} = \frac{F(Q) - F(P)}{t}$$

$$\approx \text{Directional derivative of } F \text{ in the direction of } \overrightarrow{PQ}.$$

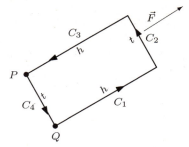

Figure 20.4: Path R used to find curl $\vec{F}$ at P

(b) Since $\vec{F} = F(x, y)\vec{T} = F(x, y)a\vec{i} + F(x, y)b\vec{j}$, with a, b constant, we have

$$\text{curl }\vec{F} = (bF_x - aF_y)\vec{k}.$$

Also $\vec{T} \times \vec{k} = (a\vec{i} + b\vec{j}) \times \vec{k} = b\vec{i} - a\vec{j}$, so

$$bF_x - aF_y = (\text{grad}F) \cdot (b\vec{i} - a\vec{j}) = \text{grad}F \cdot ((a\vec{i} + b\vec{j}) \times \vec{k}) = F_{\vec{T} \times \vec{k}},$$

where $F_{\vec{T} \times \vec{k}}$ is the directional derivative of F in the direction of the unit vector $\vec{T} \times \vec{k}$, which is perpendicular to $\vec{F}$. The right-hand rule applied to $\vec{T} \times \vec{k}$ shows that $\vec{T} \times \vec{k}$ is obtained by a *clockwise* rotation of $\vec{T}$ through $90°$.

Solutions for Section 20.4

Exercises

1. To calculate $\int_C \vec{F} \cdot d\vec{r}$ directly, we compute the integral along each of the sides C_1, C_2, C_3 in Figure 20.5. Now C_1 is parameterized by

$$x(t) = t, \quad y(t) = 0, \quad z(t) = 0 \quad \text{for } 0 \leq t \leq 5, \quad \text{so } r'(t) = \vec{i}.$$

Similarly, C_2 is parameterized by

$$x(t) = 5, \quad y(t) = t, \quad z(t) = 0 \quad \text{for } 0 \leq t \leq 5, \quad \text{so } \vec{r}\,'(t) = \vec{j}.$$

Also, C_3 is parameterized by

$$x(t) = 5 - t, \quad y(t) = 5 - t, \quad z(t) = 0 \quad \text{for } 0 \leq t \leq 5, \quad \text{so } \vec{r}\,'(t) = -\vec{i} - \vec{j}.$$

Thus

$$\int_C \vec{F} \cdot d\vec{r} = \int_{C_1} \vec{F} \cdot d\vec{r} + \int_{C_2} \vec{F} \cdot d\vec{r} + \int_{C_3} \vec{F} \cdot d\vec{r}$$

$$= \int_0^5 t(\vec{i} + \vec{j}) \cdot \vec{i}\, dt + \int_0^5 (5 - t)(\vec{i} + \vec{j}) \cdot \vec{j}\, dt + \int_0^5 ((5 - t) - (5 - t))(\vec{i} + \vec{j}) \cdot (-\vec{i} - \vec{j})\, dt$$

$$= \int_0^5 t\, dt + \int_0^5 5 - t\, dt + \int_0^5 0\, dt = \int_0^5 5\, dt = 25.$$

To calculate $\int_C \vec{F} \cdot d\vec{r}$ using Stokes' Theorem, we find

$$\text{curl }\vec{F} = \begin{vmatrix} \vec{i} & \vec{j} & \vec{k} \\ \frac{\partial}{\partial x} & \frac{\partial}{\partial y} & \frac{\partial}{\partial z} \\ x - y + z & x - y + z & 0 \end{vmatrix} = -\vec{i} + \vec{j} + (1 - (-1))\vec{k} = -\vec{i} + \vec{j} + 2\vec{k}.$$

For Stokes' Theorem, the triangular region S in Figure 20.5 is oriented upward, so $d\vec{A} = \vec{k}\, dx\, dy$. Thus

$$\int_C \vec{F} \cdot d\vec{r} = \int_S \text{curl}\, \vec{F} \cdot d\vec{A} = \int_S (-\vec{i} + \vec{j} + 2\vec{k}) \cdot \vec{k}\, dx\, dy$$

$$= \int_S 2\, dx\, dy = 2 \cdot \text{Area of triangle} = 2 \cdot \frac{1}{2} \cdot 5 \cdot 5 = 25.$$

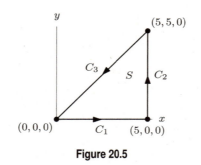

Figure 20.5

5. The boundary of S is C, the circle $x^2 + y^2 = 1$, $z = 0$, oriented counterclockwise and parameterized in polar coordinates by

$$\vec{r}(\theta) = \cos\theta\vec{i} + \sin\theta\vec{j}, \quad 0 \leq \theta \leq 2\pi,$$

so,

$$\vec{r}'(\theta) = -\sin\theta\vec{i} + \cos\theta\vec{j}.$$

Hence

$$\int_C \vec{F} \cdot d\vec{r} = \int_0^{2\pi} (\sin\theta\vec{i} + 0\vec{j} + \cos\theta\vec{k}) \cdot (-\sin\theta\vec{i} + \cos\theta\vec{j} + 0\vec{k})d\theta$$

$$= \int_0^{2\pi} -\sin^2\theta d\theta = -\pi.$$

Now consider the integral $\int_S \text{curl}\, \vec{F} \cdot d\vec{A}$. Here $\text{curl}\, \vec{F} = -\vec{i} - \vec{j} - \vec{k}$ and the area vector $d\vec{A}$, oriented upward, is given by

$$d\vec{A} = 2x\vec{i} + 2y\vec{j} + \vec{k}\, dxdy.$$

If R is the disk $x^2 + y^2 \leq 1$, then we have

$$\int_S \text{curl}\, \vec{F} \cdot d\vec{A} = \int_R (-\vec{i} - \vec{j} - \vec{k}) \cdot (2x\vec{i} + 2y\vec{j} + \vec{k})dxdy.$$

Converting to polar coordinates gives:

$$\int_S \text{curl}\, \vec{F} \cdot d\vec{A} = \int_0^{2\pi} \int_0^1 (-\vec{i} - \vec{j} - \vec{k}) \cdot (2r\cos\theta\vec{i} + 2r\sin\theta\vec{j} + \vec{k})rdrd\theta$$

$$= \int_0^{2\pi} \int_0^1 (-2r\cos\theta - 2r\sin\theta - 1)rdrd\theta$$

$$= \int_0^{2\pi} \left(\frac{2}{3}(-\cos\theta - \sin\theta) - \frac{1}{2}\right) d\theta$$

$$= -\pi.$$

Thus, we confirm that

$$\int_C \vec{F} \cdot d\vec{r} = \int_S \text{curl}\, \vec{F} \cdot d\vec{A}.$$

9. The circulation is the line integral $\int_C \vec{F} \cdot d\vec{r}$ which can be evaluated directly by parameterizing the circle, C. Or, since C is the boundary of a flat disk S, we can use Stokes' Theorem:

$$\int_C \vec{F} \cdot d\vec{r} = \int_S \text{curl}\, \vec{F} \cdot d\vec{A}$$

where S is the disk $x^2 + y^2 \leq 1$, $z = 2$ and is oriented upward (using the right hand rule). Then $\text{curl}\, \vec{F} = -y\vec{i} - x\vec{j} + \vec{k}$ and the unit normal to S is $\vec{k}$. So

$$\int_S \text{curl}\, \vec{F} \cdot d\vec{A} = \int_S (-y\vec{i} - x\vec{j} + \vec{k}) \cdot \vec{k}\, dx dy$$

$$= \int_S 1\, dx dy$$

$$= \text{Area of } S = \pi$$

Problems

13. Since

$$\text{curl}\, \vec{F} = \begin{vmatrix} \vec{i} & \vec{j} & \vec{k} \\ \frac{\partial}{\partial x} & \frac{\partial}{\partial y} & \frac{\partial}{\partial z} \\ y & -x & y-x \end{vmatrix} = \frac{\partial}{\partial y}(y-x)\vec{i} - \frac{\partial}{\partial x}(y-x)\vec{j} + \left(\frac{\partial}{\partial x}(-x) - \frac{\partial}{\partial y}(y)\right)\vec{k} = \vec{i} + \vec{j} - 2\vec{k},$$

writing S for the disk in the plane enclosed by the circle, Stokes' Theorem gives

$$\int_C \vec{F} \cdot d\vec{r} = \int_S \text{curl}\, \vec{F} \cdot d\vec{A} = \int_S (\vec{i} + \vec{j} - 2\vec{k}) \cdot d\vec{A}.$$

Now $d\vec{A} = \vec{n}\, dA$, where $\vec{n}$ is the unit vector perpendicular to the plane, so

$$\vec{n} = \frac{1}{\sqrt{3}}(\vec{i} + \vec{j} + \vec{k}).$$

Thus

$$\int_C \vec{F}\, d\vec{r} = \int_S (\vec{i} + \vec{j} - 2\vec{k}) \cdot \frac{\vec{i} + \vec{j} + \vec{k}}{\sqrt{3}}\, dA = \int_S \frac{0}{\sqrt{3}}\, dA = 0.$$

17. (a) The equation of the rim, C, is $x^2 + y^2 = 9$, $z = 2$. This is a circle of radius 3 centered on the z-axis, and lying in the plane $z = 2$.

(b) Use Stokes' Theorem, with C oriented clockwise when viewed from above:

$$\int_S \text{curl}(-y\vec{i} + x\vec{j} + z\vec{k}) \cdot d\vec{A} = \int_C (-y\vec{i} + x\vec{j} + z\vec{k}) \cdot d\vec{r}.$$

Since C is horizontal, the $\vec{k}$ component does not contribute to the integral. The remaining vector field, $-y\vec{i} + x\vec{j}$, is tangent to C, of constant magnitude $\| - y\vec{i} + x\vec{j}\| = 3$ on C, and points in the opposite direction to the orientation. Thus

$$\int_S \text{curl}(-y\vec{i} + x\vec{j} + z\vec{k}) \cdot d\vec{A} = \int_C (-y\vec{i} + x\vec{j}) \cdot d\vec{r} = -3 \cdot \text{ Length of curve} = -3 \cdot 2\pi 3 = -18\pi.$$

21. (a) We have

$$\text{curl}\, \vec{F} = \begin{vmatrix} \vec{i} & \vec{j} & \vec{k} \\ \frac{\partial}{\partial x} & \frac{\partial}{\partial y} & \frac{\partial}{\partial z} \\ y & z & x \end{vmatrix} = \vec{i}(-1) - \vec{j}(1) + \vec{k}(-1) = -\vec{i} - \vec{j} - \vec{k}.$$

(b)   (i) Using Stokes' Theorem, with S representing the disk inside the circle, oriented upward, we have

$$\int_C \vec{F} \cdot d\vec{r} = \int_S \text{curl } \vec{F} \cdot d\vec{A} = \int_S (-\vec{i} - \vec{j} - \vec{k}) \cdot \vec{k} \, dA = - \text{ Area of disk } = -4\pi.$$

(ii) This is a right triangle in the plane $x = 2$; it has height 5 and base length 3. Using Stokes' Theorem, with S representing the triangle, oriented toward the origin (in the direction $-\vec{i}$), we have

$$\int_C \vec{F} \cdot d\vec{r} = \int_S \text{curl } \vec{F} \cdot d\vec{A} = \int_S (-\vec{i} - \vec{j} - \vec{k}) \cdot (-\vec{i} \, dA) = \int_S dA = \text{ Area of triangle } = \frac{1}{2} \cdot 3 \cdot 5 = \frac{15}{2}.$$

25. (a) It appears that div $\vec{F} < 0$, and div $\vec{G} < 0$; div $\vec{G}$ is larger in magnitude (more negative) if the scales are the same.
   (b) curl $\vec{F}$ and curl $\vec{G}$ both appear to be zero at the origin (and elsewhere).
   (c) Yes, the cylinder with axis along the z-axis will have negative flux through it (ends parallel to xy-plane).
   (d) Same as part(c).
   (e) No, you cannot draw a closed curve around the origin such that $\vec{F}$ has a non-zero circulation around it because curl is zero. By Stokes' theorem, circulation equals the integral of the curl over the surface bounded by the curve.
   (f) Same as part(e)

29. (a) $\vec{F}$ has only $\vec{i}$ and $\vec{j}$ components, and they do not depend on z. Thus $\vec{F}$ is everywhere parallel to the xy-plane, and takes the same values for every value of z.
   (b) We have

$$\text{curl } \vec{F} = \begin{vmatrix} \vec{i} & \vec{j} & \vec{k} \\ \frac{\partial}{\partial x} & \frac{\partial}{\partial y} & \frac{\partial}{\partial z} \\ F_1(x, y) & F_2(x, y) & 0 \end{vmatrix} = \left(\frac{\partial F_2}{\partial x} - \frac{\partial F_1}{\partial y} \right) \vec{k}.$$

   (c) Since C is in the xy-plane, oriented counterclockwise when viewed from above, for an area element $d\vec{A}$ in S, we have $d\vec{A} = \vec{k} \, dx \, dy$. Thus Stokes' Theorem says

$$\int_C \vec{F} \cdot d\vec{r} = \int_S \text{curl } \vec{F} \cdot d\vec{A} = \int_S \left(\frac{\partial F_2}{\partial x} - \frac{\partial F_1}{\partial y} \right) \vec{k} \cdot \vec{k} \, dx \, dy = \int_S \left(\frac{\partial F_2}{\partial x} - \frac{\partial F_1}{\partial y} \right) dx \, dy.$$

   (d) Green's Theorem.

Solutions for Section 20.5

Exercises

1. Since curl $\vec{F} = \vec{0}$ and $\vec{F}$ is defined everywhere, we know by the curl test that $\vec{F}$ is a gradient field. In fact, $\vec{F} = \text{grad} f$, where $f(x, y, z) = xyz + yz^2$, so f is a potential function for $\vec{F}$.

Problems

5. Let $\vec{v} = a\vec{i} + b\vec{j} + c\vec{k}$ and try

$$\vec{F} = \vec{v} \times \vec{r} = (a\vec{i} + b\vec{j} + c\vec{k}) \times (x\vec{i} + y\vec{j} + z\vec{k}) = (bz - cy)\vec{i} + (cx - az)\vec{j} + (ay - bx)\vec{k}.$$

Then

$$\text{curl } \vec{F} = \begin{vmatrix} \vec{i} & \vec{j} & \vec{k} \\ \frac{\partial}{\partial x} & \frac{\partial}{\partial y} & \frac{\partial}{\partial z} \\ bz - cy & cx - az & ay - bx \end{vmatrix} = 2a\vec{i} + 2b\vec{j} + 2c\vec{k}.$$

Taking $a = 1$, $b = -\frac{3}{2}$, $c = 2$ gives curl $\vec{F} = 2\vec{i} - 3\vec{j} + 4\vec{k}$, so the desired vector field is $\vec{F} = (-\frac{3}{2}z - 2y)\vec{i} + (2x - z)\vec{j} + (y + \frac{3}{2}x)\vec{k}$.

9. Since div $\vec{G} = 2x + 2y + 2z \neq 0$, there is not a vector potential for $\vec{G}$.

13. (a) Using the product rule from Problem 28 on page 430, we find

$$\text{curl}\,\vec{E} = \text{curl}\left(\frac{\vec{r}}{\|\vec{r}\|^p}\right) = \frac{1}{\|\vec{r}\|^p}\,\text{curl}\,\vec{r} + \text{grad}\left(\frac{1}{\|\vec{r}\|^p}\right) \times \vec{r}.$$

Now $\text{curl}\,\vec{r} = \vec{0}$ and $\text{grad}\left(\frac{1}{\|\vec{r}\|^p}\right)$ is parallel to $\vec{r}$, so both terms are zero. Thus $\text{curl}\,\vec{E} = \vec{0}$.

(b) The domain of $\vec{E}$ is 3-space minus the origin if $p > 0$, and it is all of 3-space if $p \le 0$.

(c) Both domains have the property that any closed curve can be contracted to a point without hitting the origin, so $\vec{E}$ satisfies the curl test for all p. Since $\vec{E}$ has constant magnitude r^{1-p} on the sphere of radius r centered at the origin, and is parallel to the outward normal at every point of the sphere, the sphere must be a level surface of the potential function ϕ, that is, ϕ is a function of r alone. Further, since $\|\vec{E}\| = r^{1-p}$, a good guess is

$$\phi(r) = \int r^{1-p}\,dr,$$

that is,

$$\phi(r) = \begin{cases} \frac{r^{2-p}}{2-p} & \text{if } p \ne 2 \\ \ln r & \text{if } p = 2. \end{cases}$$

You can check that this is indeed a potential function for $\vec{E}$ by checking that $\text{grad}\,\phi = \vec{E}$.

Solutions for Chapter 20 Review

Exercises

1. Scalar. $\text{div}((2\sin(xy) + \tan z)\vec{i} + (\tan y)\vec{j} + (e^{x^2+y^2})\vec{k}) = 2y\cos(xy) + 1/\cos^2 y$.

5. We have

$$\text{div}\,\vec{F} = \frac{\partial}{\partial x}(\cos x) + \frac{\partial}{\partial y}(e^y) + \frac{\partial}{\partial z}(x+y+z) = -\sin x + e^y + 1$$

$$\text{curl}\,\vec{F} = \begin{vmatrix} \vec{i} & \vec{j} & \vec{k} \\ \frac{\partial}{\partial x} & \frac{\partial}{\partial y} & \frac{\partial}{\partial z} \\ \cos x & e^y & x+y+z \end{vmatrix} = \vec{i} - \vec{j}.$$

So $\vec{F}$ is not solenoidal and not irrotational.

9. C_2, C_3, C_4, C_6, since line integrals around C_1 and C_5 are clearly nonzero. You can see directly that $\int_{C_2} \vec{F} \cdot d\vec{r}$ and $\int_{C_6} \vec{F} \cdot d\vec{r}$ are zero, because C_2 and C_6 are perpendicular to their fields at every point.

13. (a) Direct method:

$$\text{curl}\,\vec{F} = \begin{vmatrix} \vec{i} & \vec{j} & \vec{k} \\ \frac{\partial}{\partial x} & \frac{\partial}{\partial y} & \frac{\partial}{\partial z} \\ 0 & xz & -xy \end{vmatrix} = -2x\vec{i} - (-y)\vec{j} + z\vec{k} = -2x\vec{i} + y\vec{j} + z\vec{k}.$$

On the surface, $d\vec{A}$ has no $\vec{i}$-component, so the $\vec{i}$-component of $\text{curl}\,\vec{F}$ does not contribute to the flux. Thus

$$\int_S \text{curl}\,\vec{F} \cdot d\vec{A} = \int_S (y\vec{j} + z\vec{k}) \cdot d\vec{A}.$$

Since $y\vec{j} + z\vec{k}$ is perpendicular to S and $\|y\vec{j} + z\vec{k}\| = \sqrt{y^2 + z^2} = \sqrt{5}$ on S, we have

$$\int_S \text{curl}\,\vec{F} \cdot d\vec{A} = \sqrt{5} \cdot \text{Area of } S = \sqrt{5} \cdot 2\pi\sqrt{5} \cdot 3 = 30\pi.$$

(b) Using Stokes' theorem, we replace the flux integral by two line integrals around the circular boundaries, C_1 and C_2, of S. See Figure 20.6.

$$\int_S \text{curl}\,\vec{F} \cdot ds = \int_{C_1} \vec{F} \cdot dr + \int_{C_2} \vec{F} \cdot d\vec{r}.$$

On C_1, the left boundary, $x = 0$, so $\vec{F} = \vec{0}$, and therefore $\int_{C_1} \vec{F} \cdot d\vec{r} = 0$. On C_2, the right boundary, $x = 3$, so $\vec{F} = 3z\vec{j} - 3y\vec{k}$. This vector field has $||\vec{F}|| = \sqrt{(3z)^2 + (-3y)^2} = \sqrt{9(z^2 + y^2)} = 3\sqrt{5}$. and $\vec{F}$ is tangent to the boundary C_2 and pointing in the same direction as C_2. Thus

$$\int_{C_2} \vec{F} \cdot d\vec{r} = ||\vec{F}|| \cdot \text{Length of } C_2 = 3\sqrt{5} \cdot 2\pi\sqrt{5} = 30\pi.$$

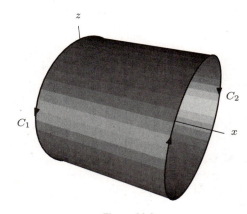

Figure 20.6

17. Since $\text{div}\,\vec{F} = 3x^2 + 3y^2$, using cylindrical coordinates to calculate the triple integral gives

$$\int_S \vec{F} \cdot d\vec{A} = \int_{\substack{\text{Interior} \\ \text{of cylinder}}} (3x^2 + 3y^2)\, dV = 3\int_0^{2\pi} \int_0^5 \int_0^2 r^2 \cdot r\, dr\, dz\, d\theta = 3 \cdot 2\pi \cdot 5 \frac{r^4}{4}\bigg|_0^2 = 120\pi.$$

21. If C is the rectangular path around the rectangle, traversed counterclockwise when viewed from above, Stokes' Theorem gives

$$\int_S \text{curl}\,\vec{F} \cdot d\vec{A} = \int_C \vec{F} \cdot d\vec{r}.$$

The $\vec{k}$ component of $\vec{F}$ does not contribute to the line integral, and the $\vec{j}$ component contributes to the line integral only along the segments of the curve parallel to the y-axis. Thus, if we break the line integral into four parts

$$\int_S \text{curl}\,\vec{F} \cdot d\vec{A} = \int_{(0,0)}^{(3,0)} \vec{F} \cdot d\vec{r} + \int_{(3,0)}^{(3,2)} \vec{F} \cdot d\vec{r} + \int_{(3,2)}^{(0,2)} \vec{F} \cdot d\vec{r} + \int_{(0,2)}^{(0,0)} \vec{F} \cdot d\vec{r},$$

we see that the first and third integrals are zero, and we can replace $\vec{F}$ by its $\vec{j}$ component in the other two

$$\int_S \text{curl}\,\vec{F} \cdot d\vec{A} = \int_{(3,0)}^{(3,2)} (x + 7)\vec{j} \cdot d\vec{r} + \int_{(0,2)}^{(0,0)} (x + 7)\vec{j} \cdot d\vec{r}.$$

Now $x = 3$ in the first integral and $x = 0$ in the second integral and the variable of integration is y in both, so

$$\int_S \text{curl}\,\vec{F} \cdot d\vec{A} = \int_0^2 10\, dy + \int_2^0 7\, dy = 20 - 14 = 6.$$

Problems

25. (a) The cube is in Figure 20.7. The vector field is parallel to the x-axis and zero on the yz-plane. Thus the only contribution to the flux is from S_2. On S_2, $x = c$, the normal is outward. Since $\vec{F}$ is constant on S_2, the flux through face S_2 is

$$\int_{S_2} \vec{F} \cdot d\vec{A} = \vec{F} \cdot \vec{A}\, _{S_2}$$
$$= c\vec{i} \cdot c^2 \vec{i}$$
$$= c^3.$$

Thus, total flux through box $= c^3$.

(b) Using the geometric definition of divergence

$$\operatorname{div} \vec{F} = \lim_{c \to 0} \left(\frac{\text{Flux through box}}{\text{Volume of box}} \right)$$
$$= \lim_{c \to 0} \left(\frac{c^3}{c^3} \right)$$
$$= 1$$

(c) Using partial derivatives,

$$\operatorname{div} \vec{F} = \frac{\partial}{\partial x}(x) + \frac{\partial}{\partial y}(0) + \frac{\partial}{\partial z}(0) = 1 + 0 + 0 = 1.$$

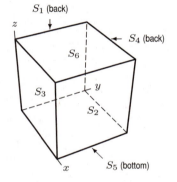

Figure 20.7

29. We close the cylinder, S, by adding the circular disk, S_1, at the top, $z = 3$. The surface $S + S_1$ is oriented outward, so S_1 is oriented upward. Applying the Divergence Theorem to the closed surface $S + S_1$ enclosing the region W, we have

$$\int_{S+S_1} \vec{F} \cdot d\vec{A} = \int_W \operatorname{div} \vec{F}\, dV$$

$$\int_S \vec{F} \cdot d\vec{A} + \int_{S_1} \vec{F} \cdot d\vec{A} = \int \operatorname{div} \vec{F}\, dV.$$

Since

$$\operatorname{div} \vec{F} = \operatorname{div}(z^2 \vec{i} + x^2 \vec{j} + 5\vec{k}) = 0,$$

we have

$$\int_S \vec{F} \cdot d\vec{A} = -\int_{S_1} \vec{F} \cdot d\vec{A}.$$

Only the $\vec{k}$-component of $\vec{F}$ contributes to the flux through S_1, and $d\vec{A} = \vec{k}\, dx\, dy$ on S_1, so

$$\int_S \vec{F} \cdot d\vec{A} = -\int_{S_1} (z^2 \vec{i} + x^2 \vec{j} + 5\vec{k}) \cdot \vec{k}\, dx\, dy = -\int_{S_1} 5\, dx\, dy = -5 \cdot \text{Area of } S_1 = -5\pi(\sqrt{2})^2 = -10\pi.$$

33. We close the cylinder, S, by adding the circular disk, S_1, at the top, $z = 3$. The surface $S + S_1$ is oriented outward, so S_1 is oriented upward. Applying the Divergence Theorem to the closed surface $S + S_1$ enclosing the region W, we have

$$\int_{S+S_1} \vec{F} \cdot d\vec{A} = \int_W \text{div} \, \vec{F} \, dV$$

$$\int_S \vec{F} \cdot d\vec{A} + \int_{S_1} \vec{F} \cdot d\vec{A} = \int \text{div} \, \vec{F} \, dV.$$

Since

$$\text{div} \, \vec{F} = \text{div}(x^3\vec{i} + y^3\vec{j} + \vec{k}) = 3x^2 + 3y^2,$$

we have

$$\int_S \vec{F} \cdot d\vec{A} = \int_W (3x^2 + 3y^2) \, dV - \int_{S_1} \vec{F} \cdot d\vec{A}.$$

To find the integral over W, we use cylindrical coordinates. For the integral over S_1, we use the fact that $d\vec{A} = \vec{k} \, dx \, dy$, so only the $\vec{k}$-component of $\vec{F}$ contributes to the flux.

$$\int_S \vec{F} \cdot d\vec{A} = \int_0^{2\pi} \int_{-3}^{3} \int_0^{\sqrt{2}} 3r^2 \cdot r \, dr \, dz \, d\theta - \int_{S_1} (x^3\vec{i} + y^3\vec{j} + \vec{k}) \cdot \vec{k} \, dx \, dy$$

$$= \theta \Big|_0^{2\pi} \, z \Big|_{-3}^{3} \, \frac{3}{4}r^4 \Big|_0^{\sqrt{2}} - \int_{S_1} dx \, dy$$

$$= 2\pi \cdot 6 \cdot \frac{3}{4}(\sqrt{2})^4 - \text{Area of } S_1$$

$$= 36\pi - \pi(\sqrt{2})^2 = 34\pi.$$

37. Since

$$\text{div}(3x\vec{i} + 4y\vec{j} + xy\vec{k}) = 3 + 4 + 0 = 7,$$

we calculate the flux using the Divergence Theorem:

$$\text{Flux} = \int_S (3x\vec{i} + 4y\vec{j} + xy\vec{k}) \cdot d\vec{A} = \int_W 7 \, dV = 7 \cdot \text{Volume of box} = 7 \cdot 3 \cdot 5 \cdot 2 = 210.$$

41. We use Stokes' Theorem. Since

$$\text{curl} \, \vec{F} = \begin{vmatrix} \vec{i} & \vec{j} & \vec{k} \\ \frac{\partial}{\partial x} & \frac{\partial}{\partial y} & \frac{\partial}{\partial z} \\ x+y & y+2z & z+3x \end{vmatrix} = -2\vec{i} - 3\vec{j} - \vec{k},$$

if S is the interior of the square, then

$$\int_C \vec{F} \cdot d\vec{r} = \int_S \text{curl} \, \vec{F} \cdot d\vec{A} = \int_S (-2\vec{i} - 3\vec{j} - \vec{k}) \cdot d\vec{A}.$$

Since the area vector of S is $49\vec{j}$, we have

$$\int_C \vec{F} \cdot d\vec{r} = \int_S (-2\vec{i} - 3\vec{j} - \vec{k}) \cdot d\vec{A} = -3\vec{j} \cdot 49\vec{j} = -147.$$

45. By the Divergence Theorem, since $\text{div} \, \vec{F} = 0$, the flux through the cone equals the flux upward through the disk $r \leq 4$ in the plane $z = 4$. The area vector of the disk is $\vec{A} = \pi 4^2 \vec{k}$. Since the flux is negative, $\vec{F} = -c(\vec{i} + \vec{k})$ with $c > 0$. Thus

$$\text{Flux} = \vec{F} \cdot \vec{A} = -c(\vec{i} + \vec{k}) \cdot \pi 4^2 \vec{k} = -16\pi c = -7.$$

so

$$c = \frac{7}{16\pi} \quad \text{and} \quad \vec{F} = -\frac{7}{16\pi}(\vec{i} + \vec{k}).$$

49. Since div $\vec{F} = 1 + 1 + 1 = 3$, the flux through the closed cylinder, S_1, with interior W, is

$$\int_{S_1} \vec{F} \cdot d\vec{A} = \int_W 3 \, dV = 3 \cdot \text{Volume of cylinder} = 3\pi.$$

With the base of the cylinder oriented downward and the top of the cylinder oriented upward,

$$\int_S \vec{F} \cdot d\vec{A} = \int_{S_1} \vec{F} \cdot d\vec{A} - \int_{\text{Base}} \vec{F} \cdot d\vec{A} - \int_{\text{Top}} \vec{F} \cdot d\vec{A}.$$

Since $\vec{F}$ is parallel to the base, the flux through the base is 0. The flux through the top is contributed entirely by the $\vec{k}$ component. Since $z = 1$, we have

$$\text{Flux through top} = \int_{\text{Top}} \vec{F} \cdot d\vec{A} = \int_{\text{Top}} (x\vec{i} + y\vec{j} + \vec{k}) \cdot d\vec{A} = \int_{\text{Top}} \vec{k} \cdot d\vec{A} = \pi.$$

Thus

$$\int_S \vec{F} \cdot d\vec{A} = 3\pi - \pi = 2\pi.$$

53. (a) Can be computed. If W is the interior of the sphere, by the Divergence Theorem, we have

$$\int_S \vec{F} \cdot d\vec{A} = \int_W \text{div} \, \vec{F} \, dV = 4 \cdot \text{Volume of sphere} = 4 \cdot \frac{4}{3}\pi \cdot 2^3 = \frac{128\pi}{3}.$$

(b) Cannot be computed.

(c) Can be computed. Use the fact that $\text{div}(\text{curl } \vec{F}) = 0$. If W is the inside of the sphere, then by the Divergence Theorem,

$$\int_S \text{curl } \vec{F} \cdot d\vec{A} = \int_W \text{div}(\text{curl } \vec{F}) dV = \int_W 0 \, dV = 0$$

57. (a) We have

$$\text{curl } \vec{F} = \begin{vmatrix} \vec{i} & \vec{j} & \vec{k} \\ \dfrac{\partial}{\partial x} & \dfrac{\partial}{\partial y} & \dfrac{\partial}{\partial z} \\ \dfrac{-y}{x^2 + y^2} & \dfrac{x}{x^2 + y^2} & 0 \end{vmatrix} = 0\vec{i} + 0\vec{j} + \left(\dfrac{\partial}{\partial x}\left(\dfrac{x}{x^2 + y^2} \right) + \dfrac{\partial}{\partial y}\left(\dfrac{y}{x^2 + y^2} \right) \right)\vec{k}.$$

Since

$$\frac{\partial}{\partial x}\left(\frac{x}{x^2 + y^2} \right) = \frac{1}{x^2 + y^2} - \frac{x(2x)}{(x^2 + y^2)^2} = \frac{x^2 + y^2 - 2x^2}{(x^2 + y^2)^2} = \frac{y^2 - x^2}{(x^2 + y^2)^2}$$

and similarly $\dfrac{\partial}{\partial y}\left(\dfrac{y}{x^2 + y^2} \right) = \dfrac{x^2 - y^2}{(x^2 + y^2)^2}$, we have, provided $x^2 + y^2 \neq 0$,

$$\text{curl } \vec{F} = \vec{0}.$$

The domain of curl $\vec{F}$ is all points in 3-space except the z-axis.

(b) On C_1, the unit circle $x^2 + y^2 = 1$ in the xy-plane, the vector field $\vec{F}$ is tangent to the circle and $||\vec{F}|| = 1$. Thus

$$\text{Circulation} = \int_{C_1} \vec{F} \cdot d\vec{r} = ||\vec{F}|| \cdot \text{Perimeter of circle} = 2\pi.$$

Note that Stokes' Theorem cannot be used to calculate this circulation since the z-axis pierces any surface which has this circle as boundary.

(c) Consider the disk $(x - 3)^2 + y^2 \leq 1$ in the plane $z = 4$. This disk has C_2 as boundary and curl $\vec{F} = \vec{0}$ everywhere on this disk. Thus, by Stokes' Theorem $\int_{C_2} \vec{F} \cdot d\vec{r} = 0$.

(d) The square S has an interior region which is pierced by the z-axis, so we cannot use Stokes' Theorem. We consider the region, D, between the circle C_1 and the square S. See Figure 20.8. Stokes' Theorem applies to the region D, provided C_1 is oriented clockwise. Then we have

$$\int_{C_1 \text{(clockwise)}} \vec{F} \cdot d\vec{r} + \int_S \vec{F} \cdot d\vec{r} = \int_D \text{curl} \, \vec{F} \cdot d\vec{A} = 0.$$

Thus,

$$\int_S \vec{F} \cdot d\vec{r} = -\int_{C_1 \text{(clockwise)}} \vec{F} \cdot d\vec{r} = \int_{C_1 \text{(counterclockwise)}} \vec{F} \cdot d\vec{r} = 2\pi.$$

(e) If a simple closed curve goes around the z-axis, then it contains a circle C of the form $x^2 + y^2 = a^2$. The circulation around C is 2π or -2π, depending on its orientation. A calculation similar to that in part (d) then shows that the circulation around the curve is 2π or -2π, again depending on its orientation. If the closed curve does not go around the z-axis, then curl $\vec{F} = \vec{0}$ everywhere on its interior and the circulation is zero.

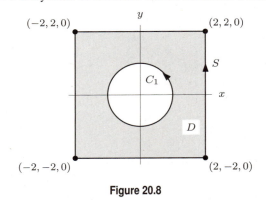

Figure 20.8

CHECK YOUR UNDERSTANDING

1. True. By Stokes' Theorem, the circulation of $\vec{F}$ around C is the flux of curl $\vec{F}$ through the flat disc S in the xy-plane enclosed by the circle. An area element for S is $d\vec{A} = \pm\vec{k}\,dA$, where the sign depends on the orientation of the circle. Since curl $\vec{F}$ is perpendicular to the z-axis, curl $\vec{F} \cdot d\vec{A} = \pm(\text{curl}\,\vec{F} \cdot \vec{k})dA = 0$, so the flux of curl $\vec{F}$ through S is zero, hence the circulation of $\vec{F}$ around C is zero.

5. True. div $\vec{F}$ is a scalar whose value depends on the point at which it is calculated.

9. False. The divergence is a scalar function that gives flux density at a point.

13. True. curl $\vec{F}$ is a vector whose value depends on the point at which it is calculated.

17. False. Since $\vec{F}$ can be written $\vec{F}(x, y, z) = x\vec{i} + y\vec{j} + z\vec{k}$, the divergence of $\vec{F}$ is 3.

21. True. By the Divergence theorem, $\int_S \vec{F} \cdot d\vec{A} = -\int_W \text{div}\,\vec{F}\,dV$, where W is the solid interior of S and the negative sign is due to the inward orientation of S. Since div $\vec{F} = 0$, we have $\int_S \vec{F} \cdot d\vec{A} = 0$.

25. True. The boundary of the cube W consists of six squares, but four of them are parallel to the xz or yz-planes and so contribute zero flux for this particular vector field. The only two surfaces of the boundary with nonzero flux are S_1 and S_2, which are parallel to the xy-plane.

29. True. The circulation density is obtained by dividing the circulation around a circle C (a scalar) by the area enclosed by C (also a scalar), in the limit as the area tends to zero.

33. False. The left-hand side of the equation does not make sense. The quantity $(\vec{F} \cdot \vec{G})$ is a scalar, so we cannot compute the curl of it.

37. False. For example, take $\vec{F} = z\vec{i}$ and $\vec{G} = x\vec{j}$. Then $\vec{F} \times \vec{G} = xz\vec{k}$, and $\text{curl}(\vec{F} \times \vec{G}) = -z\vec{j}$. However, $(\text{curl}\vec{F}) \times (\text{curl}\vec{G}) = \vec{j} \times \vec{k} = \vec{i}$.

41. True. By Stokes' theorem, both flux integrals are equal to the line integral $\int_C \vec{F} \cdot d\vec{r}$, where C is the circle $x^2 + y^2 = 1$, oriented counterclockwise when viewed from the positive z-axis.

45. True. By Stokes' theorem, $\int_S \text{curl}\,\vec{F} \cdot d\vec{A} = \int_C \vec{F} \cdot d\vec{r}$, where C is one or more closed curves that form the boundary of S. Since $\vec{F}$ is a gradient field, its line integral over any closed curve will be zero.

APPENDIX

Solutions for Section A

1. The graph is

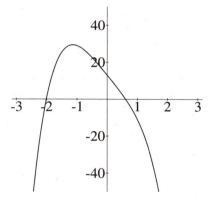

(a) The range appears to be $y \leq 30$.
(b) The function has two zeros.

5. The largest root is at about 2.5.

9. Using a graphing calculator, we see that when x is around 0.45, the graphs intersect.

13. (a) Only one real zero, at about $x = -1.15$.
(b) Three real zeros: at $x = 1$, and at about $x = 1.41$ and $x = -1.41$.

17. (a) Since f is continuous, there must be one zero between $\theta = 1.4$ and $\theta = 1.6$, and another between $\theta = 1.6$ and $\theta = 1.8$. These are the only clear cases. We might also want to investigate the interval $0.6 \leq \theta \leq 0.8$ since $f(\theta)$ takes on values close to zero on at least part of this interval. Now, $\theta = 0.7$ is in this interval, and $f(0.7) = -0.01 < 0$, so f changes sign twice between $\theta = 0.6$ and $\theta = 0.8$ and hence has two zeros on this interval (assuming f is not *really* wiggly here, which it's not). There are a total of 4 zeros.
(b) As an example, we find the zero of f between $\theta = 0.6$ and $\theta = 0.7$. $f(0.65)$ is positive; $f(0.66)$ is negative. So this zero is contained in $[0.65, 0.66]$. The other zeros are contained in the intervals $[0.72, 0.73]$, $[1.43, 1.44]$, and $[1.7, 1.71]$.
(c) You've found all the zeros. A picture will confirm this; see Figure A.1.

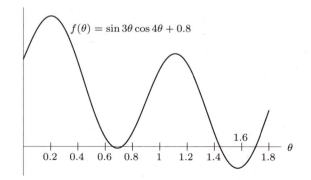

$f(\theta) = \sin 3\theta \cos 4\theta + 0.8$

Figure A.1

21.

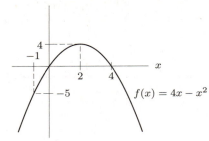

$$f(x) = 4x - x^2$$

Bounded and $-5 \leq f(x) \leq 4$.

Solutions for Section B

1. $2e^{i\pi/2}$

5. $0e^{i\theta}$, for any θ.

9. $-3 - 4i$

13. $\frac{1}{4} - \frac{9i}{8}$

17. $5^3(\cos \frac{3\pi}{2} + i \sin \frac{3\pi}{2}) = -125i$

21. One value of $\sqrt[3]{i}$ is $\sqrt[3]{e^{i\frac{\pi}{2}}} = (e^{i\frac{\pi}{2}})^{\frac{1}{3}} = e^{i\frac{\pi}{6}} = \cos \frac{\pi}{6} + i \sin \frac{\pi}{6} = \frac{\sqrt{3}}{2} + \frac{i}{2}$

25. One value of $(-4 + 4i)^{2/3}$ is $[\sqrt{32}e^{(i3\pi/4)}]^{(2/3)} = (\sqrt{32})^{2/3}e^{(i\pi/2)} = 2^{5/3}\cos \frac{\pi}{2} + i2^{5/3}\sin \frac{\pi}{2} = 2i\sqrt[3]{4}$

29. We have

$$i^{-1} = \frac{1}{i} = \frac{1}{i} \cdot \frac{i}{i} = -i,$$

$$i^{-2} = \frac{1}{i^2} = -1,$$

$$i^{-3} = \frac{1}{i^3} = \frac{1}{-i} \cdot \frac{i}{i} = i,$$

$$i^{-4} = \frac{1}{i^4} = 1.$$

The pattern is

$$i^n = \begin{cases} -i & n = -1, -5, -9, \cdots \\ -1 & n = -2, -6, -10, \cdots \\ i & n = -3, -7, -11, \cdots \\ 1 & n = -4, -8, -12, \cdots . \end{cases}$$

Since 36 is a multiple of 4, we know $i^{-36} = 1$.
Since $41 = 4 \cdot 10 + 1$, we know $i^{-41} = -i$.

33. To confirm that $z = \dfrac{a + bi}{c + di}$, we calculate the product

$$z(c + di) = \left(\frac{ac + bd}{c^2 + d^2} = \frac{bc - ad}{c^2 + d^2}i\right)(c + di)$$

$$= \frac{ac^2 + bcd - bcd + ad^2 + (bc^2 - acd + acd + bd^2)i}{c^2 + d^2}$$

$$= \frac{a(c^2 + d^2) + b(c^2 + d^2)i}{c^2 + d^2} = a + bi.$$

37. True, since $\sqrt{a}$ is real for all $a \geq 0$.

41. True. We can write any nonzero complex number z as $re^{i\beta}$, where r and β are real numbers with $r > 0$. Since $r > 0$, we can write $r = e^c$ for some real number c. Therefore, $z = re^{i\beta} = e^c e^{i\beta} = e^{c+i\beta} = e^w$ where $w = c + i\beta$ is a complex number.

45. Using Euler's formula, we have:

$$e^{i(2\theta)} = \cos 2\theta + i \sin 2\theta$$

On the other hand,

$$e^{i(2\theta)} = \left(e^{i\theta}\right)^2 = (\cos\theta + i\sin\theta)^2 = (\cos^2\theta - \sin^2\theta) + i(2\cos\theta\sin\theta)$$

Equating real parts, we find

$$\cos 2\theta = \cos^2\theta - \sin^2\theta.$$

49. Replacing θ by $(x + y)$ in the formula for $\sin\theta$:

$$
\begin{aligned}
\sin(x + y) &= \frac{1}{2i}\left(e^{i(x+y)} - e^{-i(x+y)}\right) = \frac{1}{2i}\left(e^{ix}e^{iy} - e^{-ix}e^{-iy}\right) \\
&= \frac{1}{2i}\left((\cos x + i\sin x)(\cos y + i\sin y) - (\cos(-x) + i\sin(-x))(\cos(-y) + i\sin(-y))\right) \\
&= \frac{1}{2i}\left((\cos x + i\sin x)(\cos y + i\sin y) - (\cos x - i\sin x)(\cos y - i\sin y)\right) \\
&= \sin x \cos y + \cos x \sin y.
\end{aligned}
$$

Solutions for Section C

1. (a) $f'(x) = 3x^2 + 6x + 3 = 3(x + 1)^2$. Thus $f'(x) > 0$ everywhere except at $x = -1$, so it is increasing everywhere except perhaps at $x = -1$. The function is in fact increasing at $x = -1$ since $f(x) > f(-1)$ for $x > -1$, and $f(x) < f(-1)$ for $x < -1$.

(b) The original equation can have at most one root, since it can only pass through the x-axis once if it never decreases. It must have one root, since $f(0) = -6$ and $f(1) = 1$.

(c) The root is in the interval $[0, 1]$, since $f(0) < 0 < f(1)$.

(d) Let $x_0 = 1$.

$$
\begin{aligned}
x_0 &= 1 \\
x_1 &= 1 - \frac{f(1)}{f'(1)} = 1 - \frac{1}{12} = \frac{11}{12} \approx 0.917 \\
x_2 &= \frac{11}{12} - \frac{f\left(\frac{11}{12}\right)}{f'\left(\frac{11}{12}\right)} \approx 0.913 \\
x_3 &= 0.913 - \frac{f(0.913)}{f'(0.913)} \approx 0.913.
\end{aligned}
$$

Since the digits repeat, they should be accurate. Thus $x \approx 0.913$.

5. Let $f(x) = \sin x - 1 + x$; we want to find all zeros of f, because $f(x) = 0$ implies $\sin x = 1 - x$.

Graphing $\sin x$ and $1 - x$ in Figure C.2, we see that $f(x)$ has one solution at $x \approx \frac{1}{2}$.

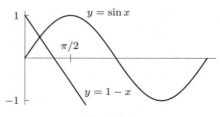

Figure C.2

Letting $x_0 = 0.5$, and using Newton's method, we have $f'(x) = \cos x + 1$, so that

$$x_1 = 0.5 - \frac{\sin(0.5) - 1 + 0.5}{\cos(0.5) + 1} \approx 0.511,$$

$$x_2 = 0.511 - \frac{\sin(0.511) - 1 + 0.511}{\cos(0.511) + 1} \approx 0.511.$$

Thus $\sin x = 1 - x$ has one solution at $x \approx 0.511$.

9. Let $f(x) = \ln x - \frac{1}{x}$, so $f'(x) = \frac{1}{x} + \frac{1}{x^2}$.
Now use Newton's method with an initial guess of $x_0 = 2$.

$$x_1 = 2 - \frac{\ln 2 - \frac{1}{2}}{\frac{1}{2} + \frac{1}{4}} \approx 1.7425,$$

$$x_2 \approx 1.763,$$

$$x_3 \approx 1.763.$$

Thus $x \approx 1.763$ is a solution. Since $f'(x) > 0$ for positive x, f is increasing: it must be the only solution.

Solutions for Section D

Exercises

1. The magnitude is $\|3\vec{i}\| = \sqrt{3^2 + 0^2} = 3$.
The angle of $3\vec{i}$ is 0 because the vector lies along the positive x-axis.

5. $2\vec{v} + \vec{w} = (2 - 2)\vec{i} + (4 + 3)\vec{j} = 7\vec{j}$.

9. Two vectors have opposite direction if one is a negative scalar multiple of the other. Since

$$5\vec{j} = \frac{-5}{6}(-6\vec{j})$$

the vectors $5\vec{j}$ and $-6\vec{j}$ have opposite direction. Similarly, $-6\vec{j}$ and $\sqrt{2}\vec{j}$ have opposite direction.

13. Scalar multiplication by 2 doubles the magnitude of a vector without changing its direction. Thus, the vector is $2(4\vec{i} - 3\vec{j}) = 8\vec{i} - 6\vec{j}$.

17. In components, the vector from $(7, 7)$ to $(9, 11)$ is $(9 - 7)\vec{i} + (11 - 9)\vec{j} = 2\vec{i} + 2\vec{j}$.
In components, the vector from $(8, 10)$ to $(10, 12)$ is $(10 - 8)\vec{i} + (12 - 10)\vec{j} = 2\vec{i} + 2\vec{j}$.
The two vectors are equal.

21. The velocity is $\vec{v}(t) = e^t\vec{i} + (1/(1 + t))\vec{j}$. When $t = 0$, the velocity vector is $\vec{v} = \vec{i} + \vec{j}$.
The speed is $\|\vec{v}\| = \sqrt{1^2 + 1^2} = \sqrt{2}$.
The acceleration is $\vec{a}(t) = e^t\vec{i} - 1/((1 + t)^2)\vec{j}$. When $t = 0$, the acceleration vector is $\vec{a} = \vec{i} - \vec{j}$.

Notes

Notes

Notes

Notes

Notes

Notes

Notes

Notes

Notes

Notes

Notes

Notes